Organic
Chemistry I

by
Frank Pellegrini, Ph.D.

INCORPORATED
LINCOLN, NEBRASKA 68501

The Cliffs Notes logo, the names "Cliffs," "Cliffs Notes," and "Cliffs Quick Review," and the black and yellow diagonal stripe cover design are all registered trademarks belonging to Cliffs Notes, Inc., and may not be used in whole or in part without written permission.

Cover photograph by Philip Habib/Tony Stone Images

FIRST EDITION

© Copyright 1997 by Cliffs Notes, Inc.

All Rights Reserved
Printed in U.S.A.

ISBN 0-8220-5326-8

Few words in the English language strike fear in the hearts of college students like the words *organic chemistry*. This fear is based more on myth than fact. Hundreds of thousands of students have successfully completed and even enjoyed the beauty of organic chemistry. What makes the difference is the learning style that is employed.

Students who appreciate organic chemistry invariably learn the concepts and apply them to the problems rather than attempt to memorize everything. The key to success in organic chemistry is to memorize the least and apply this knowledge to the fullest. You have already entered upon this pathway to success with the purchase of this Cliffs Quick Review. This text is not a treatise on organic chemistry; rather it is a concise presentation of all the important concepts, principles, and reactions of basic organic chemistry. The text is written not for the expert but for the novice who wishes to become an expert. An understanding of the topics covered in this text will allow you to read and understand the topics presented in all organic textbooks. But best of all, grasping these fundamentals will demystify the topic of organic chemistry, remove the anxiety, and make learning a pleasurable experience.

Volume I of this series covers the hydrocarbons and stereochemistry. In studying these topics, you will learn almost all of the important intermediates, types of mechanisms, and three-dimensional arrangements needed to be successful in learning organic chemistry. These concepts will be applied and further developed in volume II, in which you will study all the remaining classes of compounds.

I wish you many happy and productive hours exploring the beauty of organic chemistry.

Frank Pellegrini, Ph.D.
State University of New York
Farmingdale, NY 11735

CONTENTS

Atomic Structure

The concept of the atom was created by early Greek philosophers who believed that all matter was composed of indivisible particles. They called these particles *atomos,* meaning "uncuttable." It wasn't until the early nineteenth century that John Dalton formulated a theory based on scientific investigation that characterized the nature of atoms. Further discoveries in the nineteenth and twentieth centuries led to the knowledge that atoms possess an internal structure of smaller subatomic particles.

Subatomic particles. The major **subatomic particles** were found to be protons, electrons, and neutrons. **Protons** are positively charged particles that have weight. **Electrons** are negatively charged particles of little weight, while **neutrons** are just slightly heavier than protons but have no charge. Investigations revealed that protons and neutrons are located in the **central core,** or **nucleus,** of the atom, while electrons exist outside of the nucleus in areas of high probability called **orbits,** or **shells.** Orbits are further divided into more precise regions of electron probability called **orbitals,** or **subshells.**

Niels Bohr proposed the concept of the **solar-system atom,** in which the nucleus of the atom is like the sun and the electrons are like the planets, revolving in circular orbits. The farther an orbit is from the nucleus, the larger the orbit becomes and the more electrons it can hold.

Because all atoms are electrically neutral, the number of protons and electrons must be equal. Neutrons add weight but no charge to an atom, so additional neutrons do not change an element but merely convert it to one of its isotopic forms. The **atomic number** (Z) of an atom is equal to the number of protons in the nucleus or the number of electrons in its orbits. The **atomic mass** (A) is equal to the sum of the protons and neutrons in the atom. (A proton and neutron each have a mass of 1 atomic mass unit, while an electron has virtually no mass.)

Atoms are capable of both losing and gaining electrons to achieve a stable state. If an atom loses one or more electrons, it becomes a positively charged ion called a **cation.** If an atom gains one or more electrons, it becomes a negatively charged ion called an **anion.** The charge on an **ion** is equal to the number of electrons lost or gained.

Orbits and orbitals. Electrons fill orbits in an organized fashion based on energy factors. The order of electron fill-in, called the **aufbau buildup,** is 1s, 2s, 2p, 3s, 3p, 4s, . . . , where the numerals represent the principal quantum number of the orbit, and the lowercase letters represent the orbitals within a given orbit. The numbering begins with 1 for the orbit closest to the nucleus of the atom. The lower the orbit number, the smaller the orbit size and fewer electrons the orbit can hold.

The first principal orbit is large enough to hold just two electrons in an s orbital. The second principal orbit is large enough to contain one s and three p orbitals, while the third principal orbit, which is larger still, contains an s orbital, three p orbitals, and five d orbitals. When electrons are added to **equivalent orbitals,** which are orbitals of the same principal level and type, one electron must occupy each equivalent orbital before any of these orbitals can contain two electrons. Thus carbon, $Z = 6$, has six electrons distributed in these orbitals:

$$1s^2 2s^2 2p_x{}^1 2p_y{}^1$$

The orbitals can also be shown in the following fashion. In this diagram, the arrows represent electrons. Notice that single electrons are filling the 2p orbitals one at a time and not pairing first in $2p_x$.

$$1s \qquad 2s \qquad 2p_x \quad 2p_y \quad 2p_z$$

For two electrons to occupy the *same* orbital, they must have **opposite spins,** or paired spins, which generate orbital stability by creating opposite magnetic poles. *Between* equivalent orbitals, the spins of the electrons must be **parallel,** that is, spinning in the same direction, for the orbitals to be stable. Parallel spins create the same magnetic pole, causing repulsion between the orbitals. This repulsion gives the orbitals maximum separation and the greatest stability.

Orbitals within a given orbit have different shapes and sizes. The *s* orbitals are spherical, while the *p* orbitals are hourglass shaped. The *s* orbital is smaller than the *p* orbital.

Lewis Structures

In many instances, only the **valence electrons**—also called the **outer-shell electrons** or **bonding electrons** (because of their location in the atom and their reactivity)—are of interest to chemists. In such cases, it is advantageous to draw the Lewis structure of the atom or molecule. In a **Lewis structure** (also known as an **electron dot structure**), the entire atom, with the exception of the valence electrons, is represented by the symbol of the element, and the valence electrons are represented by dots. Thus, the Lewis structure of carbon ($Z = 6$) is

$$\cdot \overset{\displaystyle \cdot}{\underset{\displaystyle \cdot}{C}} \cdot$$

The letter C, the symbol for carbon, represents the carbon nucleus of six protons and six neutrons and the two $1s$ electrons. The four outer-shell electrons, two $2s$ and two $2p$ electrons, are represented by the dots.

Ionic Bonding

Ionic bonding occurs when electrons transfer between atoms, with a concurrent formation of ions. The electrostatic attraction between newly formed cations and anions is the heart of the ionic bond.

$$Na\cdot + \cdot\ddot{\underset{\cdot\cdot}{C}}l: \longrightarrow Na^+ + :\ddot{\underset{\cdot\cdot}{C}}l:^-$$

transfer   electrostatic
                                  attraction
                                between ions

By losing and gaining electrons, both the sodium atom and the chlorine atom acquire stability by achieving an octet of valence electrons. An **octet** of electrons is eight electrons, the number found in the outermost level of the low-atomic-weight noble gases. Through the loss of an electron, sodium becomes **isoelectronic** (having the same number and configuration of electrons) with the inert gas neon. By gaining an electron, the chlorine atom becomes isoelectronic with the inert gas argon. Ionic bonds form mainly between atoms of groups IA and IIA and atoms of groups VIA and VIIA of the periodic table.

Covalent Bonding and Electronegativity

Covalent bonds form when atoms share electrons. This sharing allows each atom to achieve its octet of electrons and greater stability. Methane, CH_4, the simplest organic compound, contains covalent bonds. Carbon has four valence electrons, while hydrogen has one valence electron. By sharing these outer-shell electrons, carbon and hydrogen complete their valence shells and become more stable. The duet of electrons on the hydrogen is isoelectronic with helium and forms a complete shell.

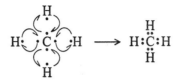

Polarity of bonds. In a **pure covalent bond,** the shared electrons are equally available to each of the atoms. This arrangement occurs only when two atoms of the same element bond with each other. Thus, the hydrogen molecule, H_2, contains a good example of a pure covalent bond.

$$H\cdot + \cdot H \longrightarrow H\!:\!H\ (H_2)$$

In most cases, the electrons in covalent bonds are not shared equally. Usually, one atom attracts the bonding electrons more strongly than does the other. This uneven attraction results in these electrons moving closer to the atom with the greater power of attraction. The resulting asymmetrical distribution of electrons makes one end of the molecule more electron rich, and it acquires a partial negative charge, while the less electron rich end acquires a partial positive charge. This difference in electron density causes the molecule to become **polar,** that is, to have a negative and a positive end.

The ability of an atom to attract electrons in a chemical bond is called the **electronegativity** of the atom. The electronegativity of an atom is related to its electron affinity and ionization energy. **Electron affinity** is the energy liberated by a gaseous atom when an electron is added to it. **Ionization energy** is the minimum amount of energy necessary to remove the most weakly bound electron from a gaseous atom.

Electronegativity level is normally measured on a scale that was created by Linus Pauling. On this scale, the more electronegative elements are the halogens, oxygen, nitrogen, and sulfur. Fluorine, a halogen, is the most electronegative with a value of 4.0, which is the highest value on the scale. The less electronegative elements are the alkali and alkaline earth metals. Of these, cesium and francium are the least electronegative at values of 0.7.

Elements with great differences in electronegativity tend to form ionic bonds. Atoms of elements with similar electronegativity tend to form covalent bonds. (Pure covalent bonds result when two atoms of the same electronegativity bond.) Intermediate differences in electronegativity between covalently bonded atoms lead to polarity in the bond. As a rule, an electronegativity difference of 2 or more on the Pauling scale between atoms leads to the formation of an ionic bond. A difference of less than 2 between atoms leads to covalent bond formation. The nearer the difference in electronegativity between atoms comes to zero, the purer the covalent bond becomes and the less polarity it has.

Carbon, with an electronegativity of 2.5, forms both low- and high-polarity covalent bonds. The electronegativity values of elements commonly found in organic molecules are given in Table 1. (See Appendix B also.)

ELECTRONEGATIVITY

Element	Value
bromine	2.8
carbon	2.5
chlorine	3.0
fluorine	4.0
hydrogen	2.1
nitrogen	3.0
oxygen	3.5
sulfur	2.5

■ Table 1 ■

Brønsted-Lowry Theory of Acids and Bases

In the early twentieth century, S. Arrhenius defined an acid as a compound that liberates hydrogen ions and a base as a compound that liberates hydroxide ions. In his acid-base theory, a neutralization is the reaction of a hydrogen ion with a hydroxide ion to form water.

$$H^+ + {}^-OH \longrightarrow H_2O$$
neutralization

The weakness of Arrhenius's theory is that it is limited to aqueous systems. A more general acid-base theory was devised by Brønsted and Lowry a couple decades later. In their theory, an acid is any compound that can donate a proton (hydrogen ion). A base is similarly defined as any substance that can accept a proton. This definition broadened the category of bases. In a Brønsted-Lowry neutralization, an acid donates a proton to a base. In the process, the original acidic molecule becomes a conjugate base; that is, it can accept a proton. Likewise, the base that accepted the proton becomes a conjugate acid, and it can donate a proton. Thus, in a Brønsted-Lowry neutralization reaction, conjugate acid-base pairs are generated.

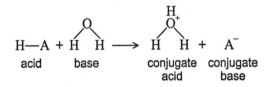

The ability of a compound to liberate protons is a measure of its strength as an acid. For a compound to easily liberate a proton, its conjugate base must be weak. Similarly, a substance that liberates protons poorly must have a conjugate base that is strong. Thus, the conjugate bases of strong mineral acids are weak, while the conjugate bases of weak inorganic and organic acids are strong.

Lewis Theory of Acids and Bases

Both the Arrhenius and Brønsted-Lowry theories of acids and bases define an acid as a hydrogen ion (proton) donor. In the Lewis theory, a base is any substance that can donate a pair of electrons to another compound. An acid then becomes any compound capable of accepting a pair of electrons from another substance. This theory greatly increases the number of chemicals considered to be acids and bases. For example, the reaction of boron trifluoride, BF_3, with dimethyl ether, CH_3OCH_3, is an acid-base reaction.

$$CH_3\overset{..}{\underset{..}{O}}CH_3 \quad + \quad BF_3 \quad \longrightarrow \quad CH_3\overset{BF_3}{\underset{..}{\overset{..}{O}}}CH_3$$

| dimethyl ether (base) | boron trifluoride (acid) | boron trifluoride dimethyletherate (neutralization product) |

Mechanisms

A **mechanism** is the series of steps that substances go through while changing from reactants to products. Each of the steps is a reaction. The step with the highest activation energy will normally be the slowest step, or **rate-determining step.** When a mechanism has been determined and proven to be correct, it allows a scientist to explain how a reaction works and to make predictions.

An illustration of a mechanism is the production of cakes at a bakery. An equation for this operation simply shows the type and amount of ingredients that are delivered to the bakery and the number and kinds of cakes that leave the bakery.

2 tn flour + 250 doz eggs + 500 lb butter + 500 lb sugar
$$\longrightarrow \text{1000 cakes of various types}$$

The mechanism tracks each step in transforming the original ingredients into the final cakes.

- flour, eggs, and water are mixed (10 minutes)

- butter, sugar, and flavorings are added (1 minute)

- batter is mixed to uniform consistency (5 minutes)

- batter is put in pans and set in oven (20 minutes)

- cakes are baked in oven (45 minutes)

- cakes are removed from oven and cooled (15 minutes)

- frosting is mixed (5 minutes)

- cakes are hand frosted and decorated (60 minutes)

These eight steps compose the mechanism for the production of the finished cakes. The slowest step, the frosting and decorating of the cakes, is the rate-determining step. If this frosting step could be modified to take less time, say, by machine decorating the cakes in thirty minutes, cake production would increase and a new rate-determining step would be in operation. In this illustration, the baking process, which requires 45 minutes, would become the rate-determining step.

Bond Rupture and Formation

Chemical reactions involve bond rupture and formation. In covalently bonded carbon molecules, for example, the bonds can be broken in two ways: symmetrically or asymmetrically. In a **symmetrical rupture,** each atom in the original covalent bond receives one electron. This type of rupture generates free radicals and is referred to as **homolytic cleavage.** In reactions, it generates free-radical mechanisms.

$$A : B \rightleftharpoons A\cdot + \cdot B$$

Asymmetrical breaking of a single covalent bond leads to ion formation and is referred to as **heterolytic cleavage.** In reactions, it generates carbocation or carbanion mechanisms. (A **carbocation** is a carbon atom bearing a positive charge; a **carbanion** is a carbon atom bearing a negative charge.)

$$A:B \rightleftharpoons A^+ + :B^-$$

The reverse reactions produce either homogenic bond formation from free radicals or heterogenic bond formation from ions.

Properties of Electrons

The Bohr model of the atom (mentioned in the previous chapter), in which electrons orbit about the nucleus, is a convenient representation. Unfortunately, it is not accurate. The results of scientific experiments suggest that electrons act more like electromagnetic waves than orbiting particles. According to a basic principle of quantum mechanics, it is impossible to know simultaneously both the exact position and momentum of an electron; this means that the trajectory of the electron cannot be precisely determined.

Waves. What can be determined is a region of space around the nucleus where there is a high probability of finding the negative charge of an atom. Mathematically, this probability distribution is similar to an equation that describes a wave. In other words, the electron distribution in an atom can be described by the mathematical formulas and physical concepts of a standing wave. A **standing wave** is basically a stationary, bound vibration—the vibration of a guitar string, for example. Viewed in slow motion, the plucking of a guitar string first displaces it a certain distance from its original position. The string rebounds to its origin and is then displaced in the opposite direction by the same distance. Figure 1 illustrates such a wave motion. The upward displacement is assigned a plus sign, while the downward one is assigned a negative sign. These signs are called **phase signs.** The point where the wave crosses the original position is called a **node,** a point of zero amplitude in the wave.

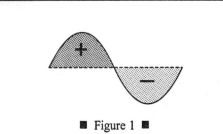

■ Figure 1 ■

Quantum numbers. In 1926, Erwin Schrödinger derived a wave equation that incorporated both the particle and wave characteristics of the electron. This equation allowed the calculation of a probability function whose solution generated four quantum numbers. The first quantum number, the **principal quantum number** (n), tells the location of the probability region relative to the nucleus. It corresponds to the orbit, or shell, designation. The second quantum number, the **angular momentum quantum number** (l), tells the shape of the probability region. It corresponds to the orbital designations. The third quantum number, the **magnetic quantum number** (m), describes the orientation of the probability region in space. The fourth quantum number, the **spin quantum number** (m_s), describes the direction of spin of the electron. The first three quantum numbers define the region in space about the nucleus of an atom where there is the highest probability of finding the electron density. This region is referred to as an **atomic orbital.**

The shapes of the two most common orbitals found in organic compounds are spherical in atomic s orbitals and hourglass shaped in atomic p orbitals. The atomic p orbitals can be oriented along the x, y, and z axes in three-dimensional space.

Linear combination of atomic orbitals. Just as waves can interact to reinforce or diminish themselves, atomic orbitals can combine to form new orbitals. This combining process is referred to as the **linear combination of atomic orbitals.** In this process, the total number of orbitals remains constant; that is, the number of new orbitals always equals the number of orbitals combined to form them. Linear combination can occur between orbitals of two different atoms, creating **molecular orbitals,** or between two orbitals of the same atom, creating **hybrid atomic orbitals.**

Molecular Orbitals

When two hydrogen atoms come together to form the hydrogen molecule, the atomic *s* orbitals of each atom are combined to form two molecular orbitals. One of these new orbitals is the result of the addition of the two atomic orbitals, while the other is created by a subtraction of these orbitals. In the addition, a reinforcement of the wave function occurs in the region between the two nuclei. Physically, this means the electron density increases in the area between the two nuclei. This increase in electron density causes a corresponding increase in the attraction of each positively charged nucleus for the negatively charged overlap region. It is this increased attraction that holds the hydrogen molecule together and creates the **bonding molecular orbital.** Because the bonding molecular orbital is generated from atomic *s* orbitals, it is called a σ (sigma) **bonding molecular orbital.**

The molecular orbital formed by the subtraction of the two wave functions has no electron density between the nuclei of the hydrogen atoms. This lack of electron density is caused by interference between the two out-of-phase wave functions. The lack of electron density between the nuclei results in the formation of a node. With no electron density between them, the hydrogen nuclei repel each other strongly, resulting in the formation of a high-energy state called an **antibonding orbital.** Because this particular antibonding orbital is created from two atomic *s* orbitals, it is referred to as a σ^* **antibonding molecular orbital.** The bonding and antibonding orbitals in the hydrogen molecule are illustrated in Figure 2. Note that the plus and minus symbols in the figure refer to wave phases and not electrical charge.

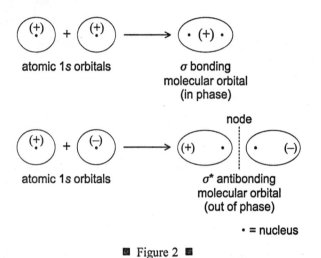

Figure 2

Like electrons in atomic orbitals, electrons in bonding orbitals must have **paired spins**; that is, the electrons must be spinning in opposite directions.

An energy diagram for the formation of the hydrogen molecule and the nonformation of a helium molecule are shown in Figure 3. The two electrons in a hydrogen molecule are paired in the lower-energy σ bonding molecular orbital. No electrons (arrows, in the figure) occupy the σ* antibonding orbital. As long as a molecule has more electrons in the bonding orbital than in the antibonding orbital, it will be stable. In fact, in most stable molecules, the antibonding orbitals are vacant. Helium molecules do not exist because no driving force causes the helium atoms to bond. They have the same number of bonding and antibonding electrons, and thus achieve no greater stability as a molecule.

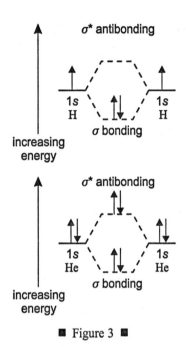

■ Figure 3 ■

The hydrogen molecule illustrates that a **σ bond** must have a heavy electron density along an imaginary line between the two nuclei. Such a density can also exist in the end-to-end overlap of atomic p orbitals. As shown in Figure 4, the heaviest electron density lies along a line between the nuclei.

atomic p orbitals   σ bonding molecular orbital
(end to end)

■ Figure 4 ■

The overlap of atomic s orbitals with hybrid atomic orbitals and the overlap of two hybrid atomic orbitals can also result in σ bonds.

The side-to-side overlap of atomic p orbitals results in high electron density above and below an imaginary line between the nuclei.

This density pattern, the **π molecular orbital,** leads to the formation of a **π (pi) bond.** This bond is much weaker than a σ bond because the repulsion between the electronically unshielded nuclei leads to poor orbital overlap.

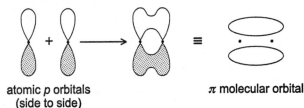

atomic *p* orbitals
(side to side)

π molecular orbital

■ Figure 5 ■

Hybridization of Atomic Orbitals

Physical studies of the simplest organic compound, methane (CH_4), have shown the following:

- all of the carbon-hydrogen bond lengths are equal
- all of the hydrogen-carbon-hydrogen bond angles are equal
- all of the bond angles are approximately 110°
- all of the bonds are covalent

The **ground state,** or unexcited state, of the carbon atom ($Z = 6$) has the following electron configuration.

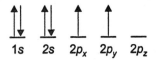

$1s$ $2s$ $2p_x$ $2p_y$ $2p_z$

Covalent bonds are formed by the sharing of electrons, so ground-state carbon cannot bond because it has only two half-filled orbitals available for bond formation. Adding energy to the system promotes a $2s$ electron to a $2p$ orbital, with the resulting generation of an excited state. The excited state has four half-filled orbitals, each capable of forming a covalent bond. However, these bonds would not all be of the same length because atomic s orbitals are shorter than atomic p orbitals.

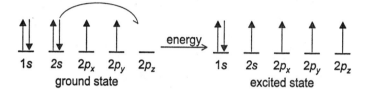

To achieve equal bond lengths, all the orbitals would have to be the same type. The creation of identical orbitals occurs in nature by a hybridization process. **Hybridization** is an internal linear combination of atomic orbitals, in which the wave functions of the atomic s and p orbitals are added together to generate new hybrid wave functions. When four atomic orbitals are added together, four hybrid orbitals form. Each of these hybrid orbitals has one part s character and three parts p character and, therefore, are called sp^3 **hybrid orbitals.**

In the hybridization process, all bond lengths become equal. Bond angles can be explained by the **valence-shell electron-pair repulsion theory (VSEPR theory).** According to this theory, electron pairs repel each other; therefore, the electron pairs that are in bonds or in lone pairs in orbitals around an atom generally separate from each other as much as possible. Thus, for methane, with four single bonds around a single carbon, the maximum angle of repulsion is the tetrahedral angle, which is 109°28", or approximately 110°.

In a similar fashion, the atomic orbitals of carbon can hybridize to form sp^2 **hybrid orbitals.** In this case, the atomic orbitals that undergo linear combination are one s and two p orbitals. This combination leads to the generation of three equivalent sp^2 hybrid orbitals. The third p orbital remains an unhybridized atomic orbital. Because the three hybrid orbitals lie in one plane, the VSEPR theory predicts that the orbitals are separated by 120° angles. The unhybridized atomic *p* orbital lies at a 90° angle to the plane. This configuration allows for the maximum separation of all orbitals.

Last, the atomic orbitals of carbon can hybridize by the linear combination of one *s* and one *p* orbital. This process forms two equivalent *sp* **hybrid orbitals.** The remaining two atomic *p* orbitals remain unhybridized. Because the two *sp* hybrid orbitals are in a plane, they must be separated by 180°. The atomic *p* orbitals exist at right angles to each other, one in the plane of the hybridized orbitals and the other at a right angle to the plane.

The type of hybrid orbital in any given carbon compound can be easily predicted with the **hybrid orbital number rule.**

$$\text{hybrid orbital number} = \sum \sigma \text{ bonds} + \text{unshared electron pairs}$$

A hybrid orbital number of 2 indicates *sp* hybridization, a value of 3 indicates sp^2 hybridization, and a value of 4 indicates sp^3 hybridization. For example, in ethene (C_2H_4), the hybrid orbital number for the carbon atoms is 3, indicating sp^2 hybridization.

$$\begin{array}{ccc} H & & H \\ | & & | \\ H-C & = & C-H \end{array}$$

All the carbon-hydrogen bonds are σ, while one bond in the double bond is σ and the other is π.

$$\text{hybrid orbital number} = 3 \ \sigma \text{ bonds} + 0 \text{ unshared electron pairs}$$
$$= 3$$

Thus, the carbons have sp^2 hybrid orbitals.

Using the hybrid orbital number rule, it can be seen that the methylcarbocation contains sp^2 hybridization, while the methylcarbanion is sp^3 hybridized.

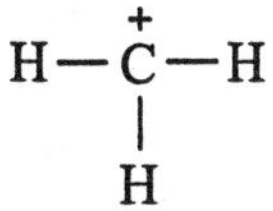

$$H - \overset{+}{\underset{|}{C}} - H$$

H

methylcarbocation
hybrid orbital number = 3

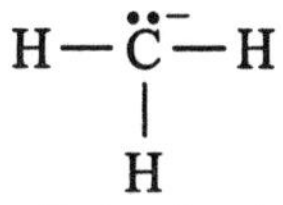

$$H - \overset{\cdot\cdot\,-}{\underset{|}{C}} - H$$

H

methylcarbanion
hybrid orbital number = 4

Three-dimensional Shapes of Molecules

The overall shape of an organic molecule is fixed by the shape of the central carbon atoms, which compose the backbone of the molecule. The shape of this backbone is determined by the types of hybrid orbitals making up the bonds between the central carbon atoms. If the central carbon atoms are sp^3 hybridized, the molecule will possess a tetrahedral shape. Central carbon atoms that are sp^2 hybridized lead to trigonal-planar shapes, while sp hybridization produces linear molecules. Three-dimensional representations of methane (sp^3 hybridization), ethene (sp^2 hybridization), and ethyne (sp hybridization) molecules are shown in Figure 6.

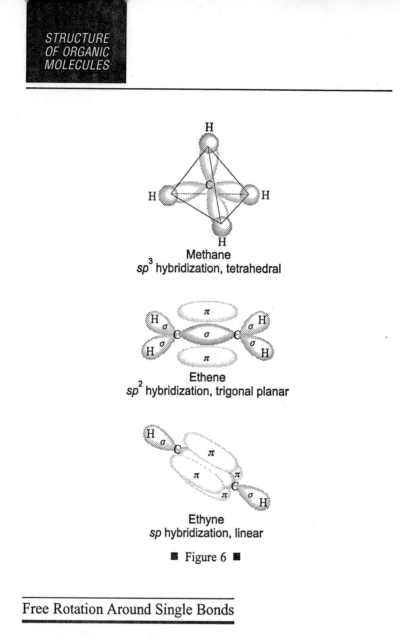

Methane
sp^3 hybridization, tetrahedral

Ethene
sp^2 hybridization, trigonal planar

Ethyne
sp hybridization, linear

■ Figure 6 ■

Free Rotation Around Single Bonds

Carbon atoms in single bonds rotate freely. Rotation can occur because the heaviest electron density in the σ bond exists along an imaginary line between two carbon nuclei. Rotation does not change this electron distribution; the bond strength remains constant throughout rotation. Because rotation is possible, the molecule can have an infinite number of conformations, and a sketch of any of them is an accurate representation of the molecule. Some conformations have slightly higher energy content because of repulsion between atoms or between groups bonded to adjacent carbons as they approach each other due to rotation. Rotation around the carbon-carbon bond in the ethane molecule is shown in Figure 7.

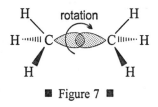

■ Figure 7 ■

Nonrotation Around Multiple Bonds

Double and triple bonds are referred to as **multiple bonds.** These bonds are composed of a σ bond and either one or two π bonds. Because the π bonds are created by the side-to-side overlap of atomic *p* orbitals, any rotation around the σ bond results in the destruction of all of the π bonds. The effect on the double bond of ethene if rotation is attempted around the σ bond is shown in Figure 8.

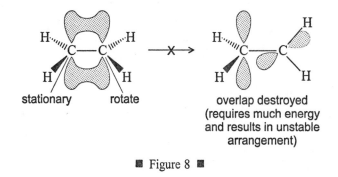

■ Figure 8 ■

Structural Isomers and Stereoisomers

Isomers are compounds with different physical and chemical properties but the same molecular formula. In organic chemistry, there are many cases of isomerism. For example, the formula C_4H_{10} represents both butane and 2-methylpropane.

$$CH_3 - CH_2 - CH_2 - CH_3$$
butane

$$CH_3 - \overset{\overset{\displaystyle CH_3}{|}}{CH} - CH_3$$
2-methylpropane

These are examples of **structural isomers,** or **constitutional isomers.** Structural isomers have the same molecular formula but a different bonding arrangement among the atoms.

    Stereoisomers have identical molecular formulas and arrangements of atoms. They differ from each other only in the spatial orientation of groups in the molecule. The simplest forms of stereoisomers are *cis* and *trans* isomers, both of which are created by the restricted rotation about a double bond or ring system. Butene, C_4H_8, exists in both *cis* and *trans* forms.

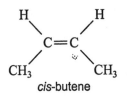
cis-butene

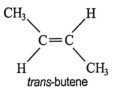
trans-butene

Hydrocarbons

Hydrocarbons comprise the simplest series of organic compounds. These compounds are composed of carbon and hydrogen atoms in single, double, and triple bonds. Depending on the type of central bond present, the hydrocarbon can be classified as an **alkane** (single bond), **alkene** (double bond), or **alkyne** (triple bond). A **diene** contains two double bonds. These are the functional groups that are reviewed in this book.

Molecular and Structural Formulas

The alkanes comprise a series of compounds that are composed of carbon and hydrogen atoms with single covalent bonds. This group of compounds comprises a homologous series with a general molecular formula of C_nH_{2n+2}, where n equals any integer.

The simplest alkane, methane, has one carbon atom and a molecular formula of CH_4. Because this compound contains only single covalent bonds, its structural formula is

methane

In longer alkane molecules, the additional carbon atoms are attached to each other by single covalent bonds. Each carbon atom is also attached to sufficient hydrogen atoms to produce a total of four single covalent bonds about itself. Thus octane, an eight-carbon alkane, has a molecular formula of C_8H_{18} and a structural formula of

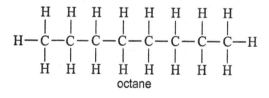

octane

Alkyl groups. When a substituent, such as a halogen or hydroxy group, bonds to an alkane molecule, one of the carbon-hydrogen bonds of the molecule is converted to a carbon-substituent bond. For example, when methane reacts with chlorine, a new compound called chloromethane (or methyl chloride) is formed. This new compound contains a CH_3 group bonded to a chlorine atom.

$$CH_4 \; + \; Cl_2 \; \xrightarrow{\text{UV light}} \; CH_3Cl \; + \; HCl$$
methane   chlorine   chloromethane   hydrogen
chloride

An alkane with a hydrogen removed from one bond is called an **alkyl group.** Alkyl groups are often represented by the letter **R,** just as halogens are often represented by the letter **X.** The methane-chlorine reaction can be generalized as

$$R-H + X_2 \; \xrightarrow{\text{UV light}} \; R-X + HX$$

Often, organic chemists use these types of representations for discussing generalized reactions.

Isomers. Alkyl groups can exist in more than one isomeric form. For example, the alkane propane has two alkyl isomers.

$$CH_3CH_2CH_2- \quad \text{and} \quad CH_3CHCH_3$$
$$|$$

These isomers are distinguished from each other by the type of carbon that has lost a hydrogen to form the alkyl group. Carbon atoms are classified as primary (1°), secondary (2°), or tertiary (3°), depending on the number of carbon atoms to which they are attached. A **primary carbon** is directly attached to only one other carbon atom. **Secondary** and **tertiary carbons** are attached to two and three other carbon atoms, respectively.

In the propane isomer diagram above, the group on the left is a primary (1°) propyl group, while the group on the right is a secondary (2°) propyl group. Butyls (alkanes with four carbons) have three isomeric groups. The structures and names of these groups are

$$CH_3CH_2CH_2CH_2-$$ 1° butyl

$$CH_3CH_2CHCH_3$$ 2° butyl
       |

$$
\begin{array}{c}
CH_3 \\
| \\
CH_3-C- \\
| \\
CH_3
\end{array}
$$ 3° butyl

Nomenclature

Although many different types of **nomenclature,** or naming systems, were employed in the past, today only the **International Union of Pure and Applied Chemistry (IUPAC) nomenclature** is acceptable for all scientific publications. In this system, a series of rules has been created that is adaptable to all classes of organic compounds. For alkanes, the following rules apply.

1. Identify the longest continuous chain of carbon atoms. The **parent name** of the alkane is the IUPAC-assigned name for the alkane of this number of carbon atoms (see Table 2). Thus, if the longest chain of carbon atoms has six carbon atoms in it, the parent name for the compound is hexane.

2. Identify the substituent groups attached to the parent chain. A **substituent group** is any atom or group that has replaced a hydrogen atom on the parent chain.

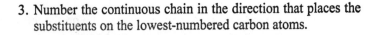

3. Number the continuous chain in the direction that places the substituents on the lowest-numbered carbon atoms.

4. Write the name of the compound. The parent name is the last part of the name. The name(s) of the substituent group(s) and the location number(s) precede the parent name. A hyphen separates the number associated with the substituent from its name. If two substituents are on the same carbon of the parent chain, the number of the carbon they are attached to is written before *each* substituent name. If the two substituents are identical, the numbers are both written before the substituent name, and the prefix "di" is added to the name. Substituent group names are placed in alphabetical order.

Applying the four nomenclature rules to the following compound

$$
\begin{array}{c}
\text{Cl} \\
|\\
\text{CH}_3\text{CHCHCH}_3 \ (b) \\
|\\
\text{CH}_2\text{CH}_3 \ (a)
\end{array}
$$

results in the name 2-chloro-3-methylpentane. Notice that the parent name comes from the longest continuous carbon chain, which begins with the carbon of the CH_3 group at the bottom of the structural formula (*a*) and goes to the carbon of the CH_3 group on the top right side of the formula (*b*). This chain contains five carbon atoms, while the straight chain of carbons from left to right contains only four carbons. Starting the numbering from the top right carbon of the CH_3 group (*b*) leads to 2,3 substitution, while numbering from the bottom right side CH_3 carbon (*a*) leads to 3,4 substitution (which is incorrect). This alkane is referred to as a **branched-chain alkane** because it contains an alkyl group off of the main chain.

Applying the IUPAC nomenclature rules to a more complex alkane molecule

$$\underset{\substack{| \\ OH}}{CH_3CHCH_2}\underset{\substack{| \\ Cl}}{CHCHCH_2CH_3}$$
$$\underset{\substack{| \\ CH_2CH_2CH_3}}{}$$

results in the name 5-chloro-2-hydroxy-4-propylheptane. Notice that the names of the substituent groups are in alphabetical order.

Finally, here is an example of a compound with a repeating substituent group.

$$\underset{\substack{| \\ Cl}}{CH_3CH}CH_2\underset{\substack{| \\ Cl}}{CH}CH_2CH_3$$

This compound is called 2,4-dichlorohexane.

IUPAC NAMES OF THE FIRST TWENTY PARENT ALKANES

Number of carbon atoms	Name	Number of carbon atoms	Name
1	methane	11	undecane
2	ethane	12	dodecane
3	propane	13	tridecane
4	butane	14	tetradecane
5	pentane	15	pentadecane
6	hexane	16	hexadecane
7	heptane	17	heptadecane
8	octane	18	octadecane
9	nonane	19	nonadecane
10	decane	20	icosane

■ Table 2 ■

Physical Properties

The alkanes can exist as gases, liquids, or solids at room temperature. The unbranched alkanes methane, ethane, propane, and butane are gases; pentane through hexadecane are liquids; the homologues larger than hexadecane are solids.

Branched alkanes normally exhibit lower boiling points than unbranched alkanes of the same carbon content. This occurs because of the greater van der Waals forces that exist between molecules of the unbranched alkanes. These forces can be dipole-dipole, dipole-induced dipole, or induced dipole-induced dipole in nature. The unbranched alkanes have greater van der Waals forces of attraction because of their greater surface areas.

Solid alkanes are normally soft, with low melting points. These characteristics are due to strong repulsive forces generated between electrons on neighboring atoms, which are in close proximity in crystalline solids. The strong repulsive forces counterbalance the weak van der Waals forces of attraction.

Finally, alkanes are almost completely insoluble in water. For alkanes to dissolve in water, the van der Waals forces of attraction between alkane molecules and water molecules would have to be greater than the dipole-dipole forces that exist between water molecules. This is *not* the case.

Natural Sources

The alkanes are isolated from natural gas and petroleum. Natural gas contains mainly methane, with smaller amounts of other low-molecular-weight alkanes. Petroleum, which is a complex mixture of many compounds, is the main source of all other alkanes. The lighter fractions are distilled from the mixture to produce the liquid alkanes, while the residue from the distillation produces the solid alkanes.

Preparations

Alkanes are rarely prepared from other types of compounds because of economic reasons. However, ignoring financial considerations, alkanes can be prepared from the following compounds:

1. Unsaturated compounds via catalytic reduction

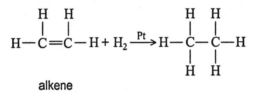

alkene

2. Alkyl halides via coupling (Wurtz reaction)

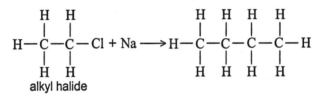

alkyl halide

3. Alkyl halides via Grignard reagent

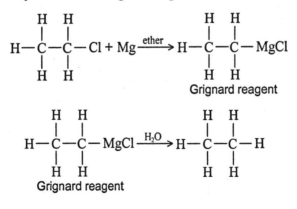

Grignard reagent

Grignard reagent

4. Alkyl halides via reduction

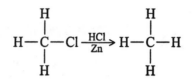

Although organic chemists refer to the above diagrams as "equations," they are not balanced. In addition, not every product formed is shown. These diagrams are really **reaction schemes.**

Alkanes do not react with most reagents for two reasons. First, carbon-carbon and carbon-hydrogen single bonds are very strong due to good orbital overlap. Second, the carbon-hydrogen bonds make alkane molecules neither acidic nor basic because the electronegativity of both elements is very similar. This similarity gives the carbon-hydrogen bond little polarity, and without polarity, proton loss is difficult. Thus, alkanes make poor acids. Likewise, a lack of nonbonded electron pairs on either the carbon or hydrogen atoms makes alkanes poor bases. However, under proper conditions, alkanes can react with halogens and oxygen.

Halogenation

The reaction of a halogen with an alkane in the presence of ultraviolet (UV) light or heat leads to the formation of a **haloalkane (alkyl halide)**. An example is the chlorination of methane.

$$
\underset{\underset{H}{\overset{H}{|}}}{H-C-H} + Cl_2 \xrightarrow[\text{or 400°C}]{\text{UV light}} \underset{\underset{H}{\overset{H}{|}}}{H-C-Cl} + HCl
$$

Experiments have shown that when the alkane and halogen reactants are not exposed to UV light or heat, the reaction does not occur. However, once a reaction is started, the light or heat source can be removed and the reaction will continue. The mechanism of the reaction explains this phenomenon.

Halogenation mechanism. In the methane molecule, the carbon-hydrogen bonds are low-polarity covalent bonds. The halogen molecule has a nonpolar covalent bond. UV light contains sufficient energy to break the weaker nonpolar chlorine-chlorine bond (~58 kcal/mole), but it has insufficient energy to break the stronger carbon-hydrogen bond (104 kcal/mole). The fracture of the chlorine molecule leads to the formation of two highly reactive chlorine free radicals (chlorine atoms). A **free radical** is an atom or group that has a single unshared electron.

$$Cl \text{:} Cl \xrightarrow{\text{UV light}} Cl \cdot + Cl \cdot$$
chlorine
free radicals

The bond that is ruptured is said to have broken in a **homolytic** fashion; that is, each of the originally bonded atoms receives one electron. This initial reaction is called the **initiation step** of the mechanism. The chlorine free radicals that form are in a high-energy state and react quickly to complete their octets and liberate energy. Once the high-energy chlorine free radicals are formed, the energy source (UV light or heat) can be removed. The energy liberated in the reaction of the free radicals with other atoms is sufficient to keep the reaction running.

When a chlorine free radical approaches a methane molecule, a homolytic fission of a carbon-hydrogen bond occurs. The chlorine free radical combines with the liberated hydrogen free radical to form hydrogen chloride and a methyl free radical. This is called a **propagation step,** a step in which both a product and a reactive species, which keeps the reaction going, are formed.

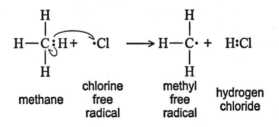

methane · chlorine free radical · methyl free radical · hydrogen chloride

A second propagation step is possible. In this step, a methyl free radical reacts with a chlorine molecule to form chloromethane and a chlorine free radical.

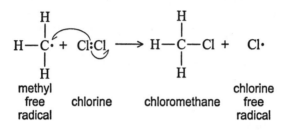

| methyl free radical | chlorine | chloromethane | chlorine free radical |

When a reaction occurs between free radicals, a product forms, but no new free radicals are formed. This type of reaction is called a **termination step** because it tends to end the reaction. There are several termination steps in the chlorination of methane.

1. A methyl free radical reacts with a chlorine free radical to form chloromethane.

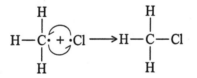

2. Two methyl free radicals react to form ethane.

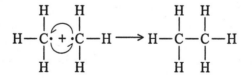

3. Two chlorine free radicals react to form a chlorine molecule.

$$Cl\underset{\smile}{\overset{\frown}{+}}Cl \longrightarrow Cl-Cl$$

To summarize, this **free-radical chain reaction** initially contains few free radicals and many molecules of reactants. As the reaction proceeds, the number of free radicals increases, while the number of reactant molecules decreases. Near the end of the reaction, many more free radicals exist than reactant molecules. At this stage of the overall reaction, termination steps become the predominant reactions. All of the halogenation mechanism reactions occur very rapidly, and the formation of the products takes only microseconds.

Kinetics and Rate

Most reactions require the addition of energy. Energy is needed for molecules to pass over the energy barriers that separate them from becoming reaction products. These energy barriers are called the **activation energy,** or **enthalpy of activation,** of the reactions.

At room temperature, most molecules have insufficient kinetic energy to overcome the activation energy barrier so a reaction can occur. The average kinetic energy of molecules can be increased by increasing their temperature. The higher the temperature, the greater the fraction of reactant molecules that have sufficient energy to pass over the activation energy barrier. Thus, the rate of a reaction increases with increasing temperature.

The rate of a reaction also depends on the number of interactions between reactant molecules. Interactions increase in solutions of greater concentrations of reactants, so a reaction rate is directly proportional to the concentration of the reactants. The proportionality constant is called the **rate constant** for the reaction. Not every collision is effective in producing bond breakage and formation. For a collision to be effective, the molecules must have sufficient energy content as well as proper alignment. If all collisions were effective, every reaction would proceed with explosive force.

Activation energy. The change in structure of each of the reactants as a reaction proceeds is very important in organic chemistry. For example, in the reaction of methane and chlorine, the molecules of each substance must "collide" with sufficient energy, and the bonds within the molecules must be rearranged for chloromethane and hydrogen chloride to be produced. As reactant molecules approach each other, old bonds are cleaved, and new bonds are formed. The cleavage of bonds requires a lot of energy, so as the reaction occurs, the reactant molecules must remain in high-energy states. When new bonds form, energy is released, and the resulting products possess less energy than the intermediates from which they were formed. When reactant molecules are at their maximum energy content (at the crest of the activation energy curve), they are said to be in a **transition state.** The energy necessary to drive the reactants to the transition state is the **activation energy** (Figure 9).

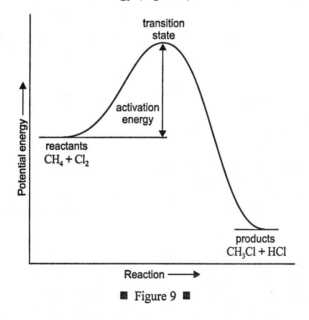

■ Figure 9 ■

Many organic reactions involve more than one step. In such cases, the reactants may proceed through one or more intermediate stages

(either stable or unstable arrangements), with corresponding transition states, before they finally form products (Figure 10).

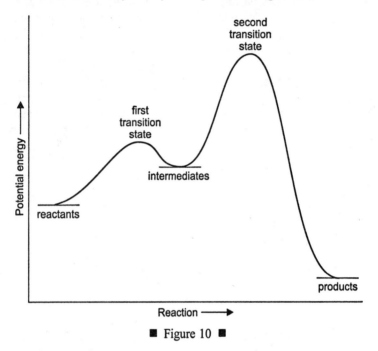

■ Figure 10 ■

The overall rate of the reaction is determined, for the most part, by the transition state of highest energy in the pathway. This transition state, which is usually the slowest step, controls the rate of reaction and is thus called the **rate-determining step** of the mechanism.

Energy of reaction. The **energy of reaction** is the difference between the total energy content of the reactants and the total energy content of the products (Figure 11). In ordinary organic reactions, the products contain less energy than the reactants, and the reactions are therefore **exothermic.** The energy of reaction has no effect on the rate of the reaction. The greater the energy of reaction, the more stable the products.

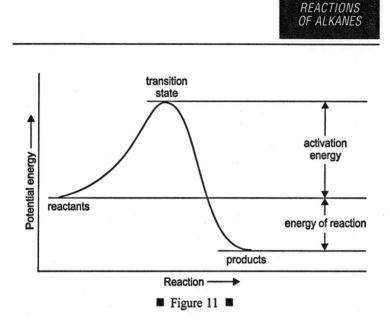

■ Figure 11 ■

Effects of temperature on rate of reaction. The rates of organic reactions approximately double with each 10°C rise in temperature. A more quantitative relationship between reaction rate and temperature is given by the Arrhenius equation

$$k = Ae^{-E_a/RT}$$

where $k =$ rate constant of the reaction

$A =$ Arrhenius factor (reflects the effectiveness as well as the number of collisions)

$e =$ natural logarithm base

$E_a =$ activation energy

$R =$ universal gas constant

$T =$ Kelvin temperature

Oxidation

Alkanes can be oxidized to carbon dioxide and water via a free-radical mechanism. The energy released when an alkane is completely oxidized is called the **heat of combustion.** For example, when propane is oxidized, the heat of combustion is 688 kilocalories per mole.

$$CH_3CH_2CH_3 + 5\,O_2 \longrightarrow 3\,CO_2 + 4\,H_2O + energy$$

In a homologous series like the straight-chain alkanes, the energy liberated during oxidation increases by approximately 157 kilocalories for each additional methylene (CH_2) unit.

Heat of combustion data is often used to assess the relative stability of isomeric hydrocarbons. Because the heat of combustion of a compound is the same as the enthalpy of that compound in its standard state, and because potential energy is comparable to enthalpy, the differences in heats of combustion between two alkanes translate directly to differences in their potential energies. The lower the potential energy of a compound, the more stable it is. In the alkanes, the more highly branched isomers are usually more stable than those that are less branched.

Molecular and Structural Formulas

The alkenes comprise a series of compounds that are composed of carbon and hydrogen atoms with at least one double bond in the carbon chain. This group of compounds comprises a homologous series with a general molecular formula of C_nH_{2n}, where n equals any integer greater than one.

The simplest alkene, ethene, has two carbon atoms and a molecular formula of C_2H_4. The structural formula for ethene is

$$\begin{array}{cc} H & H \\ | & | \\ H-C & = C-H \end{array}$$

In longer alkene chains, the additional carbon atoms are attached to each other by single covalent bonds. Each carbon atom is also attached to sufficient hydrogen atoms to produce a total of four single covalent bonds about itself. In chains with four or more carbon atoms, the double bond can be located in different positions, leading to the formation of **structural isomers.** For example, the alkene of molecular formula C_4H_8 has two isomers.

$$CH_3-CH_2-CH=CH_2 \quad \text{and} \quad CH_3-CH=CH-CH_3$$

Stereoisomers. In addition to structural isomers, alkenes also form **stereoisomers.** Because rotation around a multiple bond is restricted, groups attached to the double-bonded carbon atoms always remain in the same relative positions. These "locked" positions allow chemists to identify various isomers from the substituents' locations. For example, one structural isomer of C_5H_{10} has the following stereoisomers.

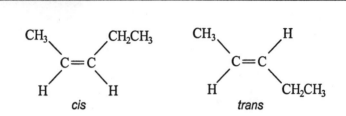

The isomer on the left, in which the two **substituents** (the methyl and ethyl groups) are on the *same side* of the double bond, is called the *cis* **isomer,** while the isomer on the right, with two nonhydrogen substituents on *opposite sides* of the double bond, is called the ***trans* isomer.**

If more than two substituents are attached to the carbon atoms of a double bond, the *cis* and *trans* system cannot be used. With such chemicals, E-Z notation is used. In the ***E-Z* system,** the molecule is first bisected vertically through the double bond. Second, the two atoms or groups on each carbon atom are ranked by atomic weight. The higher atomic weight is assigned priority. For example, in Figure 12, the carbon and chlorine atoms on the left side of the bisecting line are ranked. Chlorine has priority because it is heavier. On the right side, bromine outranks carbon. Third, the positions of the two atoms of higher rank are determined. If the two atoms are in the *cis* position, the arrangement is *Z* (for German *zusammen,* meaning "together"). If the atoms or groups are in the *trans* position, the arrangement is *E* (for German *entgegen,* meaning "opposite").

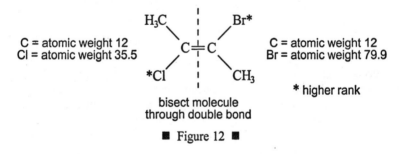

bisect molecule
through double bond

■ Figure 12 ■

The name of the chemical in Figure 12 is (*E*)-2-bromo-3-chloro-2-butene.

Unsaturation

The defining characteristic of an alkene molecule is the double bond. This bond is composed of a σ and a π covalent bond. Because π bonds are formed exclusively by the side-to-side overlap of atomic p orbitals, any rotation along the σ-bond axis requires disruption of the π-bond system. Breaking the π-bond system requires an energy input of 63 kilocalories per mole. Due to this energy barrier, free rotation about the carbon-carbon double bond is not possible. Multiple bonds between two atoms always lead to restricted rotation.

Compounds containing double bonds are said to be **unsaturated** because they are capable of reacting with hydrogen in the presence of a catalyst. The resulting compounds, which contain no multiple bonds, are said to be **saturated.** For each π bond destroyed, two σ bonds are formed, and energy is normally released. Thus chemists refer to a π bond as an **element of unsaturation.** Each element of unsaturation corresponds to two fewer hydrogen atoms than are found in the formula of the saturated compound of the same carbon-chain length.

Nomenclature

Alkenes are normally named using the IUPAC system. The rules for alkenes are similar to those used for alkanes. The following rules summarize alkene nomenclature.

1. Identify the longest continuous chain of carbon atoms that contains the carbon-carbon double bond. The parent name of the alkene comes from the IUPAC name for the alkane with the same number of carbon atoms, except the *-ane* ending is changed to *-ene* to signify the presence of a double bond. For example, if the longest continuous chain of carbon atoms containing a double bond has five carbon atoms, the compound is **a pentene.**

2. Number the carbon atoms of the longest continuous chain, starting at the end closest to the double bond. Thus,

$$CH_3-CH_2-CH=CH-CH_3$$

is numbered from right to left, placing the double bond between the second and third carbon atoms of the chain. (Numbering the chain from left to right incorrectly places the double bond between the third and fourth carbons of the chain.)

3. The position of the double bond is indicated by placing the lower of the pair of numbers assigned to the double-bonded carbon atoms in front of the name of the alkene. Thus, the compound shown in rule 2 is 2-pentene.

4. The location and name of any substituent molecule or group is indicated. For example,

$$CH_3-\overset{\overset{\displaystyle Cl}{\displaystyle |}}{C}H-CH_2-CH=CH-CH_3$$

is 5-chloro-2-hexene.

5. Finally, if the correct three-dimensional relationship is known about the groups attached to the double-bonded carbons, the *cis* or *trans* conformation label may be assigned. Thus, the complete name of the compound in rule 4 (shown differently here)

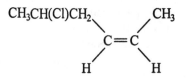

is *cis*-5-chloro-2-hexene.

Physical Properties

The physical properties of alkenes are very similar to those of alkanes. Alkenes also exist as gases, liquids, and solids at room temperature. Isomeric alkenes tend to have similar boiling points, which makes it difficult to separate them by boiling point differences.

Substituted alkenes show small dipole moments due to small electron distribution differences. These small differences allow *cis* and *trans* isomers to be distinguished from each other. The effects of substitution must be deduced for each molecule, based on the positions and the electronegativity of the atoms or groups attached to the carbon-carbon double bond. Thus, in the case of *cis*- and *trans*-2-butene, the *cis* isomer shows a dipole moment of 0.33 debye units (D), while the *trans* isomer shows a dipole moment of 0 D, due to cancellation of the electronic effects.

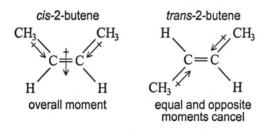

cis-2-butene · trans-2-butene

overall moment · equal and opposite moments cancel

Preparations

Alkenes are generally prepared through β elimination reactions, in which two atoms on adjacent carbon atoms are removed, resulting in the formation of a double bond.

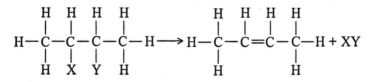

Preparations include the dehydration of alcohols, the dehydrohalogenation of alkyl halides, and the dehalogenation of alkanes.

Dehydration of alcohols. In **dehydration reactions,** a molecule of water is eliminated from an alcohol molecule by heating the alcohol in the presence of a strong mineral acid. A double bond forms between the adjacent carbon atoms that lost the hydrogen ion and hydroxide group.

$$CH_3CH_2CH_2\ddot{O}H + H_2SO_4 \xrightarrow{\text{heat}} CH_3CH{=}CH_2 + H_2O$$

The mechanism of this dehydration reaction consists of the following steps.

1. Protonation of the alcohol.

$$CH_3CH_2CH_2\ddot{O}H \xrightarrow{H^+} CH_3CH_2CH_2\overset{H}{\underset{+}{\ddot{O}}}H$$

This step is a simple acid-base reaction, which results in the formation of an **oxonium ion,** a positively charged oxygen atom.

2. Dissociation of the oxonium ion.

$$CH_3CH_2CH_2\overset{H}{\underset{+}{\ddot{O}}}H \longrightarrow CH_3CH_2\overset{+}{C}H_2 + H_2O$$

Dissociation of the oxonium ion produces a **carbocation,** which is a positively charged carbon atom and an unstable intermediate.

3. Deprotonation of the carbocation.

$$CH_3 - CH_2 - \overset{+}{C}H_2 \longrightarrow CH_3 - CH = CH_2 + H^+$$
$$\underset{H}{|}$$

The positively charged end carbon of the carbocation attracts the electrons in the overlap region that bond it to the adjacent α carbon. This electron movement makes the α carbon slightly positive, which in turn attracts the electrons in the overlap regions of all other atoms bonded to it. This results in the hydrogen on the α carbon becoming very slightly acidic and capable of being removed as a proton in an acid-base reaction.

Zaitsev rule. It may be possible in some instances to create a double bond through an alcohol dehydration reaction in which hydrogen atoms are lost from two different carbons on the carbocation. The major product is always the more highly substituted alkene, that is, the alkene with the greater number of substituents on the carbon atoms of the double bond, an observation called the **Zaitsev rule.** Thus, in the dehydration reaction of 2-butanol, the following products are formed.

$$CH_3 - CH_2 - \overset{\overset{\displaystyle OH}{|}}{CH} - CH_3 \xrightarrow{H^+, \text{ heat}}$$

$$\underset{\text{2-butene}}{CH_3 - CH = CH - CH_3} + \underset{\text{1-butene}}{CH_3 - CH_2 - CH = CH_2} + H_2O$$

The Zaitsev rule predicts that the major product is 2-butene. Notice that each carbon atom involved in the double bond of 2-butene has one methyl group attached to it. In the case of 1-butene, one carbon atom of the double bond has one substituent (the ethyl group), while the other carbon atom has no substituents.

Carbocation rearrangement. The carbocation in an alcohol dehydration may undergo **rearrangement** to form more stable arrangements. Dehydration of 2-methyl-3-pentanol, for example, leads to the production of three alkenes. The mechanism for the reaction shows that the extra compound formation is due to rearrangement of the carbocation intermediate.

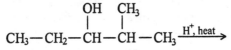

$$\underset{\text{2-methyl-3-pentanol}}{CH_3-CH_2-\underset{\underset{OH}{|}}{CH}-\underset{\underset{CH_3}{|}}{CH}-CH_3} \xrightarrow{H^+,\ heat}$$

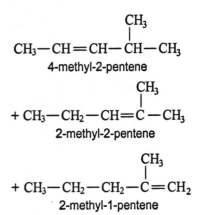

$$\underset{\text{4-methyl-2-pentene}}{CH_3-CH=CH-\underset{\underset{CH_3}{|}}{CH}-CH_3}$$

$$+\ \underset{\text{2-methyl-2-pentene}}{CH_3-CH_2-CH=\underset{\underset{CH_3}{|}}{C}-CH_3}$$

$$+\ \underset{\text{2-methyl-1-pentene}}{CH_3-CH_2-CH_2-\underset{\underset{CH_3}{|}}{C}=CH_2}$$

The 2-methyl-1-pentene molecule is formed via rearrangement of the intermediate carbocation.

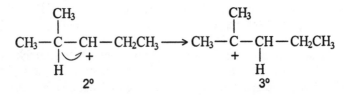

The movement of a hydride ion (H:⁻) leads to the formation of a more stable carbocation. Carbocations are classified as primary, secondary, and tertiary, as are the carbon atoms. A **primary carbocation**

has one alkyl group attached to it; a **secondary carbocation** is bonded to two alkyl groups; and a **tertiary carbocation** has three alkyl groups around it.

$$\overset{+}{CH_3CH_2} \qquad\qquad \overset{+}{CH_3CHCH_3} \qquad\qquad \overset{+}{CH_3CCH_3}$$
$$\qquad\qquad\qquad\qquad\qquad\qquad\qquad\qquad\qquad | $$
$$\qquad\qquad\qquad\qquad\qquad\qquad\qquad\qquad\quad CH_3$$

primary (1°)   secondary (2°)   tertiary (3°)

Alkyl groups theoretically have the ability to "push" electrons away from themselves. This phenomenon is called the **inductive effect.** The greater the number of alkyl groups "pushing" electrons toward a positively charged carbon atom, the more stable the intermediate carbocation will be. This increase in stability is due to the delocalization of charge density. A charge on an atom creates a stress on that atom. The more the stress is spread over the molecule, the smaller the charge density becomes on any one atom, reducing the stress. This lessening of stress makes the ion more stable. Thus, tertiary carbocations, with three alkyl groups on which to delocalize the positive charge, are more stable than secondary carbocations, which have only two alkyl groups on which to delocalize the positive charge. For the same reason, secondary carbocations are more stable than primary carbocations.

In reality, alkyl groups do not "push" electrons away from themselves, but rather they have electrons removed from them. When an atom picks up a positive charge and becomes an ion, its electronegativity changes. In the original σ bond between two carbon atoms, the location of the overlap region relative to each carbon atom is fixed in part by the electronegativity of the two atoms. With an increase in the electronegativity of one of the carbon atoms due to ion formation, the overlap region shifts closer to the more electronegative, positively charged carbon atom. This rearrangement of electron density produces a partial positive charge on the neighboring carbon. The amount of charge gained by the second carbon corresponds to the amount lost by the fully charged carbon atom. In this manner, the charge becomes delocalized over the two carbons.

Dehydrohalogenation of alkyl halides. The **dehydrohalogenation of alkyl halides,** another β elimination reaction, involves the loss of a hydrogen and a halide from an alkyl halide (RX). Dehydrohalogenation is normally accomplished by reacting the alkyl halide with a strong base, such as sodium ethoxide.

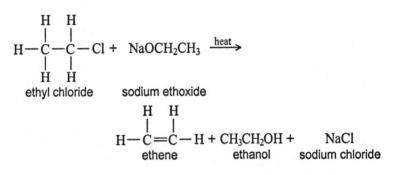

ethyl chloride   sodium ethoxide

ethene   ethanol   sodium chloride

This reaction also follows the Zaitsev rule, so in the reaction of 2-chlorobutane with sodium ethoxide, the major product is 2-butene.

$$CH_3 - CH_2CH - CH_3 + NaOCH_2CH_3 \xrightarrow{heat}$$
$$\underset{Cl}{|}$$

2-chlorobutane   sodium ethoxide

$$CH_3CH = CH - CH_3 + CH_3CH_2CH = CH_2$$

2-butene   1-butene
(major product)   (minor product)

Dehydrohalogenation reactions proceed via the following mechanism.

1. A strong base removes a slightly acidic hydrogen proton from the alkyl halide via an acid-base reaction.

2. The electrons from the broken hydrogen-carbon bond are attracted toward the slightly positive carbon atom attached to the chlorine atom. As these electrons approach the second carbon, the halogen atom breaks free, leading to the formation of

the double bond. The diagram below summarizes this mechanism.

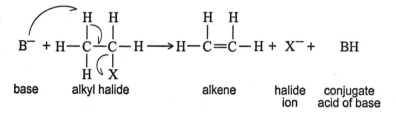

Dehalogenation. Vicinal dihalides, which are alkane molecules that contain two halogen atoms on adjacent carbon atoms, can form alkenes upon reaction with zinc.

$$H-\underset{\underset{X}{|}}{\overset{\overset{H}{|}}{C}}-\underset{\underset{X}{|}}{\overset{\overset{H}{|}}{C}}-H + Zn \longrightarrow H-\overset{\overset{H}{|}}{C}=\overset{\overset{H}{|}}{C}-H + ZnX_2$$

Electrophilic Addition Reactions

The most common reactions of the alkenes are **additions** across the double bond to form saturated molecules. Such reactions are represented by the following general equation, where X and Y represent elements in a compound that are capable of being added across the π-bond system of an alkene to form a substituted alkane.

$$X-Y + H_2C=CH_2 \longrightarrow XCH_2CH_2Y$$

Halogenation

Halogenation is the addition of halogen atoms to a π-bond system. For example, the addition of bromine to ethene produces the substituted alkane 1,2-dibromoethane.

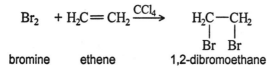

The reaction proceeds via a *trans* addition, but because of the free rotation possible around the single bond of the resulting alkane, a *trans* product cannot be isolated. If, however, the original alkene structure possesses restricted rotation due to a factor other than a double bond, a *trans*-addition product can be isolated. For instance, **ring structures** possess restricted rotation. In a ring structure, the carbon backbone is arranged so there is no beginning or ending carbon atom. If cyclohexene, a six-carbon ring that has one double bond, is halogenated, the resulting cycloalkane is *trans* substituted.

cyclohexene   *trans*-1,2-dibromocyclohexane

Mechanism and stereochemistry of halogenation. Alkenes and halogens are nonpolar molecules. However, both types of molecules, under proper conditions, can undergo induced-dipole formation, which leads to the generation of forces of attraction between the molecules.

$$CH_2 = CH_2 + Br - Br \xrightarrow{CCl_4} {}^+CH_2CH_2Br + Br^-$$

ethene   bromine   bromoethyl   bromide
                                       carbocation   ion

The bromoethyl carbocation that forms mid reaction in this example is often internally stabilized by cyclization into a three-membered ring containing a positively charged bromine atom (bromonium ion).

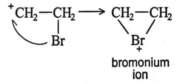

bromonium
ion

This intermediate is more stable than the corresponding linear carbocation because all the atoms have a complete octet of electrons.

The bromonium ion shares the electrons in the carbon-bromine covalent bond unevenly, with the overlap region being closer to the more electronegative bromine. This generates a partial positive charge (δ^+) on the carbon atoms of the ring. The charge delocalization stabilizes the ring structure, and the resulting partial positive charges on the carbon atoms attract the nucleophilic bromide ion.

$$Br^- \quad \overset{\delta^+}{CH_2} \overset{\delta^+}{-CH_2} \longrightarrow \overset{\overset{Br}{|}}{CH_2} - \overset{\overset{}{|}}{\underset{Br}{CH_2}}$$

The second bromide ion must approach a partially positive carbon atom from the side of the carbocation opposite where the bromonium ion attached. The reason for this is that the bromonium ion blocks access to the carbon atoms along an entire side, due to bond formation with the two carbon atoms. Such blocking is referred to as **steric hindrance.** Because of steric hindrance, only a *trans* addition is possible.

Hydrohalogenation

Unlike halogens, hydrogen halides are polarized molecules, which easily form ions. Hydrogen halides also add to alkenes by electrophilic addition.

$$CH_2 = CH_2 + HX \longrightarrow CH_3CH_2X$$
$$\text{ethene} \qquad \text{hydrogen} \qquad \text{alkyl halide}$$
$$\text{halide}$$

The addition of hydrogen halides to asymmetrically substituted alkenes leads to two products.

$$CH_3CH = CH_2 + HBr \longrightarrow CH_3CH_2CH_2Br + \overset{}{\underset{Br}{CH_3CHCH_3}}$$

| propene | hydrogen bromide | 1-bromopropane (minor product) | 2-bromopropane (major product) |

The major product is predicted by the **Markovnikov rule,** which states that when a hydrogen halide is added to an asymmetrically substituted alkene, the major product results from the addition of the hydrogen atom to the double-bonded carbon that is attached to more hydrogen atoms, while the halide ion adds to the other double-bonded carbon. This arrangement creates a more stable carbocation intermediate.

Hydrohalogenation mechanisms. The first step in the addition of a hydrogen halide to an alkene is the dissociation of the hydrogen halide.

$$HBr \rightleftharpoons H^+ + Br^-$$

The H^+ ion is attracted to the π-bond electrons of the alkene, which forms a π **complex.**

$$CH_3CH = CH_2 + H^+ \longrightarrow CH_3CH \overset{H^+}{\not=} CH_2$$
$$\text{propane} \qquad \text{proton} \qquad \pi \text{ complex}$$

The π complex then breaks, creating a σ single bond between one carbon of the double-bonded pair and the hydrogen. The carbon atom that loses a share of the π bond then becomes a carbocation. In asymmetrically substituted alkenes, two different carbocations are possible. The major product is generated from the more stable carbocation, while the minor product forms from the less stable one.

$$CH_3CH \overset{H^+}{\not=} CH_2 \longrightarrow CH_3CH_2 - \overset{+}{C}H_2 + CH_3\overset{+}{C}H - CH_3$$
$$\pi \text{ complex} \qquad 1° \text{ carbocation} \qquad 2° \text{ carbocation}$$

Thus, the major product is 2-bromopropane.

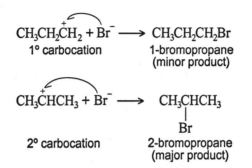

$$CH_3CH_2\overset{+}{C}H_2 + Br^- \longrightarrow CH_3CH_2CH_2Br$$

1° carbocation   1-bromopropane
(minor product)

$$CH_3\overset{+}{C}HCH_3 + Br^- \longrightarrow CH_3CHCH_3$$
$$\qquad\qquad\qquad\qquad\qquad | $$
$$\qquad\qquad\qquad\qquad\qquad Br$$

2° carbocation   2-bromopropane
(major product)

Hydrogen bromide can also be added to an alkene in an anti-Markovnikov fashion. In **anti-Markovnikov additions,** the hydrogen atom of the hydrogen halide adds to the carbon of the double bond that is bonded to *fewer* hydrogen atoms. For this to result, the reaction must proceed by a noncarbocation intermediate; thus in the presence of peroxide, the reaction proceeds via a free-radical mechanism, with the major product being generated from the more stable free radical.

$$CH_3CH = CH_2 + \quad HBr \quad \xrightarrow{\text{peroxide}}$$

propene   hydrogen
bromide

$$CH_3CH_2CH_2Br + \quad CH_3CHCH_3$$
$$\qquad\qquad\qquad\qquad\qquad\qquad | $$
$$\qquad\qquad\qquad\qquad\qquad\qquad Br$$

1-bromopropane   2-bromopropane
(major product)   (minor product)

The mechanism for this reaction starts with the generation of a bromine free radical by the reaction of hydrogen bromide with peroxide.

$$ROOR \xrightarrow{\text{light or heat}} 2RO\bullet$$

organic   alkoxy
peroxide   free radicals

$$RO\bullet + \quad HBr \quad \longrightarrow ROH + \quad Br\bullet$$

free   hydrogen   alcohol   bromine
radical   bromide   free radical

The bromine free radical adds to the alkene, forming a more stable carbon free radical.

$$CH_3CH = CH_2 + \cdot Br \longrightarrow CH_3\overset{\cdot}{C}H - CH_2Br + CH_3CH - \overset{\cdot}{C}H_2$$

$$\underset{\text{2° free radical}}{} \qquad \underset{\substack{| \\ Br \\ \text{1° free radical}}}{}$$

The secondary free radical is more stable than the primary free radical because the secondary molecule is better able to delocalize the stress placed on the carbon atom by the free-radical electron. The major product then forms from the intermediates by reacting with hydrogen bromide.

$$CH_3\overset{\cdot}{C}H - CH_2Br + HBr \longrightarrow CH_3CH_2CH_2Br + Br\cdot$$

$$\text{1-bromopropane}$$
$$\text{(major product)}$$

$$CH_3CH - \overset{\cdot}{C}H_2 + HBr \longrightarrow CH_3CHCH_3 + Br\cdot$$
$$\underset{Br}{|} \qquad\qquad\qquad\qquad \underset{Br}{|}$$

$$\text{2-bromopropane}$$
$$\text{(minor product)}$$

In all additions of hydrogen halides across carbon-carbon double bonds, the major product always comes from the more stable intermediate. In Markovnikov additions, the major product results from the more stable carbocation, while in anti-Markovnikov additions, such as the hydrogen bromide addition in the presence of peroxide, the major product results from the more stable free radical.

Hydration (Direct Addition of Water)

The addition of water to an alkene in the presence of a catalytic amount of strong acid leads to the formation of alcohols (hydroxyalkanes).

$$CH_2 = CH_2 + H_2O \xrightarrow{H^+} CH_3CH_2OH$$

This reaction proceeds via a standard carbocation mechanism and follows the Markovnikov rule. The mechanism for the addition of water to ethene follows.

1. The hydrogen ion is attracted to the π bond, which breaks to form a σ bond with one of the double-bonded carbons. The second carbon of the original double-bonded carbons becomes a carbocation.

$$CH_2 = CH_2 + H^+ \longrightarrow CH_3\overset{+}{C}H_2$$
ethene

2. An acid-base reaction occurs between the water molecule and the carbocation, forming an oxonium ion.

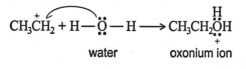
water   oxonium ion

3. The oxonium ion stabilizes by losing a hydrogen ion, with the resulting formation of an alcohol.

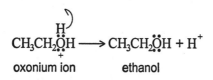
oxonium ion   ethanol

Hydroboration-Oxidation (Indirect Addition of Water)

Water can also be added to an alkene in such a way that the major product is not that predicted by the Markovnikov rule. An example of such a reaction is the indirect addition of water to an alkene via a **hydroboration-oxidation reaction.** In this reaction, a disubstituted boron hydride is added across the carbon-carbon double bond of an alkene. The resulting organoborane compound is oxidized to an alcohol by reaction with hydrogen peroxide in a basic media, such as aqueous sodium hydroxide solution.

$$CH_2 {=} CH_2 + (CH_3)_2BH \longrightarrow CH_3CH_2B(CH_3)_2$$

ethene   disubstituted   organoborane
     boron hydride

$$CH_3CH_2B(CH_3)_2 + H_2O_2 + OH^- \longrightarrow$$
$$CH_3CH_2OH + CH_3OH + B(OH)_4^- + H_2O$$

No carbocation intermediate forms during this reaction. Although the elements of water are added to an alkene, water is not a reactant; the hydrogen comes from a boron hydride molecule, and the hydroxide group comes from a peroxide molecule.

The first step in the hydroboration mechanism is the formation of the organoborane molecule from the alkene. This reaction occurs rapidly. The boron atom generally bonds to the less substituted, and thus less sterically hindered, carbon. This first step proceeds via a reaction between the disubstituted organoborane and the π bond of the alkene, followed by formation of a C–H bond via a four-center interaction. A **four-center interaction** is a reaction in which bonds between four atoms are created and broken simultaneously.

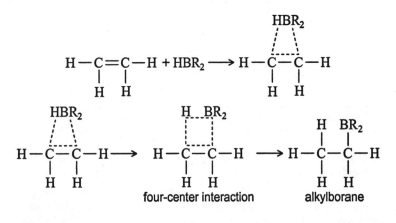

four-center interaction   alkylborane

The alkylborane then undergoes a three-stage oxidation reaction to form the alcohol. In the first step, a hydroperoxide anion, formed by the reaction of a hydroxide ion with a peroxide molecule, adds to the electron-deficient boron atom.

$$CH_3CH_2BR_2 + {}^-OOH \longrightarrow CH_3CH_2\overset{-}{B}R_2-OOH$$

alkylborane   hydroperoxide anion

This intermediate is unstable and rearranges, losing a hydroxide ion to form a borate ester.

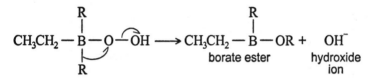

$$CH_3CH_2-\underset{R}{\overset{R}{B}}-O-OH \longrightarrow CH_3CH_2-\underset{R}{\overset{R}{B}}-OR + OH^-$$

    borate ester   hydroxide ion

The borate ester then reacts with alkaline hydrogen peroxide to produce a trialkyl borate.

$$CH_3CH_2-\overset{R}{B}-OR + OOH \longrightarrow CH_3CH_2O\overset{OR}{B}-OR + OH^-$$

borate ester   hydro-peroxide anion   trialkyl borate   hydroxide ion

Finally, the trialkyl borate is hydrolyzed (which means split by the elements of water) to alcohols and a borate ion by the aqueous hydroxide ion.

$$CH_3CH_2O\overset{OR}{B}-OR + OH^- + H_2O \longrightarrow$$

trialkyl borate   hydroxide ion   water

$$CH_3CH_2OH + ROH + B(OH)_4^-$$

ethanol   alcohol   borate ion

Catalytic Addition of Hydrogen

Hydrogenation is the addition of hydrogen to an alkene. Although this reaction is exothermic, it is very slow. The addition of a metal catalyst, such as platinum, palladium, nickel, or rhodium, greatly increases the reaction rate. Although this reaction seems simple, it is a highly complex addition. The reaction takes place in four steps.

In the first step, a hydrogen molecule reacts with the metal catalyst. This reaction breaks the σ bond between the hydrogen atoms and creates weak hydrogen-metal bonds. Next, the π bond of an alkene molecule contacts the metal catalyst. The π bond is destroyed and two weak carbon-metal single bonds are created. Finally, the weakly bound hydrogen atoms transfer one at a time from the catalyst surface to the carbon atoms of the former alkene molecule, forming an alkane. Upon formation of the two new carbon-hydrogen bonds, the alkane molecule can move away from the catalyst.

Because both of the added hydrogen atoms were bound to the surface of the catalyst, they normally approach the alkene molecule from the same side, or **face.** This approach of hydrogen atoms to the same face of an alkene molecule is called a *syn* **addition.**

$$\underline{H-H} \rightleftharpoons \underline{\overset{H}{\mid}\ \overset{H}{\mid}} \rightleftharpoons \underline{\overset{H}{\mid}\ \overset{CH_2=CH_2}{\mid}\ \overset{H}{\mid}}$$

$$\underline{\overset{H}{\mid}\ \overset{CH_2=CH_2}{\mid}\ \overset{H}{\mid}} \longrightarrow H-CH_2-CH_2-H$$

When hydrogen atoms approach alkene molecules from opposite sides, the reaction is called an *anti* **addition.** *Anti* addition most likely occurs when double-bond isomerization occurs more rapidly than the catalytic addition of the second hydrogen in the hydrogenation.

Addition of Carbenes

Carbenes are intermediates of the general formula $R_2C:$. In this configuration, the carbon atom possesses only a sextet of electrons, and is therefore highly reactive and electrophilic. Carbenes are generally prepared by reacting a haloform, such as chloroform, with a strong base, such as sodium ethoxide.

$$CHCl_3 \ + \ CH_3CH_2O^-Na^+ \longrightarrow \ Cl_2C: \ + \ Cl^-$$

chloroform	sodium ethoxide	dichlorocarbene	chloride ion

Carbene ($H_2C:$), however, is prepared by exposing diazomethane to ultraviolet light.

$$CH_2N_2 \ \xrightarrow{\text{UV light}} \ :CH_2 \ + \ N_2$$

diazomethane	carbene	nitrogen

Due to the high reactivity of carbenes, they cannot be isolated. All carbene reactions are run by generating the carbene **"in situ,"** that is, generating the carbene in the presence of a reagent with which it will immediately react. Alkenes, which are ready sources of electrons, are such reagents. When alkenes react with carbenes, three-membered rings are formed. (Ring structures are discussed in the chapter "Cyclohydrocarbons.")

$$CH_2{=}CH_2 + CH_3Cl + CH_3CH_2O^-Na^+ \longrightarrow \overset{\displaystyle Cl \quad Cl}{\underset{\displaystyle CH_2{-}CH_2}{\diagdown C \diagup}} + NaCl$$

ethene	chloro-methane	sodium ethoxide	1,1-dichloro-cyclopropane	sodium chloride

$$CH_2{=}CH_2 + \ CH_2N_2 \ \xrightarrow{\text{UV light}} \overset{\displaystyle CH_2}{\underset{\displaystyle CH_2{-}CH_2}{\diagup \diagdown}}$$

ethene	diazomethane	cyclopropane

The insertion of a carbene into a π-bond system is the most common way of preparing cyclopropanes. The addition of the methylene unit, CH_2, to the carbon-carbon double bond of the alkene is a *syn* addition.

Some chemicals, namely the **carbenoids,** behave like carbenes, even though they are not. The most common carbenoid is the Simmons-Smith reagent, a mixture of iodomethane and a zinc-copper couple. This reagent also reacts with alkenes to form a cyclopropane ring.

$$CH_3I \quad + \quad Zn(Cu) \quad \xrightarrow{\text{ether}} \quad ICH_2ZnI$$

iodomethane	zinc-copper couple	Simmons-Smith reagent (iodomethylzinc iodide)

$$CH_2{=}CH_2 + \quad ICH_2ZnI \quad \longrightarrow \quad \overset{\displaystyle CH_2}{\overset{\diagup\diagdown}{CH_2{-}CH_2}}$$

ethene	Simmons-Smith reagent	cyclopropane

The mechanisms of carbene and carbenoid reactions show the difference between the two. The mechanism for a carbene reaction is a concerted process in which all bonds are broken and formed at one time.

$$CH_2{=}CH_2 + \; {:}CH_2 \; \longrightarrow \underset{\displaystyle CH_2}{CH_2{-}CH_2} \longrightarrow \underset{\displaystyle CH_2}{CH_2{-}CH_2}$$

ethene	carbene		cyclopropane

The mechanism for the Simmons-Smith reaction also shows a concerted addition; however, a carbene is never formed.

$$CH_2{=}CH_2 + ICH_2ZnI \longrightarrow \underset{\underset{\displaystyle I \quad ZnI}{\displaystyle CH_2}}{CH_2{-}CH_2} \longrightarrow \underset{\displaystyle CH_2}{CH_2{-}CH_2} + ZnI_2$$

ethene	Simmons-Smith reagent		cyclopropane	

Epoxide Reactions

Alkenes are capable of reacting with oxygen in the presence of elemental silver to form a series of cyclic ethers called epoxides. **Epoxides** are three-atom cyclic systems in which one of the atoms is oxygen. The simplest epoxide is epoxyethane (ethylene oxide).

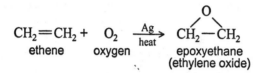

$$CH_2{=}CH_2 + \quad O_2 \quad \xrightarrow[\text{heat}]{Ag} \quad CH_2{-}CH_2$$

ethene   oxygen   epoxyethane
(ethylene oxide)

Epoxyethane belongs to a class of chemicals called **heterocyclic compounds.** These compounds are cyclic structures in which one (or more) of the ring atoms is a **hetero atom,** that is, an atom of an element other than carbon. In the laboratory, epoxyethane is prepared by reacting 1-chloro-2-hydroxyethane with a base.

$$HO{-}CH_2{-}CH_2{-}Cl \xrightarrow{\text{base}} CH_2{-}CH_2$$

1-chloro-2-hydroxyethane   epoxyethane

The mechanism for this reaction starts with the base reacting with the acidic hydrogen of the OH group.

$$HO{-}CH_2{-}CH_2{-}Cl + RO^-Na^+ \longrightarrow ClCH_2CH_2O^-Na^+ + ROH$$

1-chloro-2-hydroxyethane

The oxygen anion is then attracted to the carbon that is bonded to the chlorine atom. This carbon bears a strong partial positive charge due to the great differences in electronegativity between the carbon and chlorine atoms.

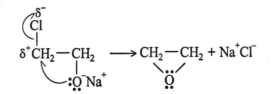

As can be seen in the structural formula above, the oxygen atom must be located *anti* to the departing chlorine atom for the reaction to occur. The overall reaction is a *syn* addition.

A third method of preparing epoxyethane is by the reaction of an alkene with peroxy acids.

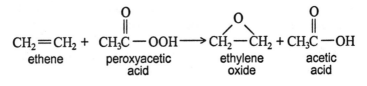

Oxidation and Cleavage Reactions

Alkenes can easily be oxidized by potassium permanganate and other oxidizing agents. What products form depend on the reaction conditions. At cold temperatures with low concentrations of oxidizing reagents, alkenes tend to form glycols.

$$H_2C = CH_2 + 1\text{–}4\% \ KMnO_4 \longrightarrow HOCH_2 - CH_2OH + MnO_2$$

| ethene | potassium permanganate | ethylene glycol | manganese dioxide |

This reaction is sometimes referred to as the **Baeyer test.** Because potassium permanganate, which is purple, is reduced to manganese dioxide, which is a brown precipitate, any water-soluble compound that produces this color change when added to cold potassium permanganate must possess double or triple bonds. This reaction involves *syn* addition, leading to a *cis*-glycol (a vicinal dihydroxy compound). A *cis*-glycol can also be produced by reacting the alkene with osmium tetroxide, OsO_4.

When more concentrated solutions of potassium permanganate and higher temperatures are employed, the glycol is further oxidized, leading to the formation of a mixture of ketones and carboxylic acids.

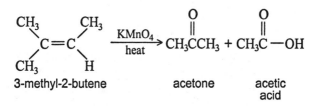

3-methyl-2-butene   acetone   acetic acid

Oxidation of alkenes by ozone leads to destruction of both the σ and π bonds of the double-bond system. This cleavage of an alkene double bond, generally accomplished in good yield, is called **ozonolysis.** The products of ozonolysis are aldehydes and ketones.

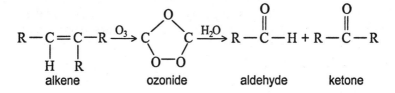

alkene   ozonide   aldehyde   ketone

This reaction is often used to find the double bond in an alkene molecule. For example, the isomers of C_4H_8 can be distinguished from one another via oxidative cleavage.

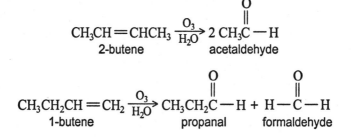

2-butene   acetaldehyde

1-butene   propanal   formaldehyde

By identifying the products of the reaction, one isomer can be distinguished from another, and the position of the bonds in the original compound can be determined.

Polymerization

Polymerization is a process by which an organic compound reacts with itself to form a high-molecular-weight compound composed of repeating units of the original compound. The polymerization of ethene by an ionic, or free-radical, reagent A–B is an example.

$$CH_2 = CH_2 \xrightarrow{AB} ACH_2CH_2 \left(\!\! CH_2CH_2 \!\!\right)_{\!n} CH_2CH_2B$$

Polymerization reactions proceed via either cationic or free-radical mechanisms. In both processes, π bonds are converted to σ bonds, and energy is liberated. Cationic polymerization is less efficient than free-radical polymerization due to the caustic nature of cation-producing reagents. An example of a cation-initiated polymerization is the reaction of ethene with sulfuric acid.

$$H_2SO_4 \rightleftharpoons H^+ + HSO_4^-$$

$$H^+ + CH_2 = CH_2 \longrightarrow CH_3 - \overset{+}{C}H_2$$

$$CH_3\overset{+}{C}H_2 + CH_2 = CH_2 \longrightarrow CH_3CH_2 - CH_2 - \overset{+}{C}H_2$$

The reaction continues and gives

$$CH_3CH_2 \left(\!\! CH_2CH_2 \!\!\right)_{\!n} CH_2CH_2^+$$

which finally reacts with HSO_4^- to create the polymer

$$CH_3CH_2 \left(\!\! CH_2CH_2 \!\!\right)_{\!n} CH_2CH_2\overset{+}{H}SO_4^-$$

The more effective free-radical polymerization can be initiated by oxygen or other free-radical compounds, such as peroxides. The free-radical polymerization of ethene by an alkoxide radical is a typical reaction.

$$RO\cdot + \overset{\frown}{CH_2 = CH_2} \longrightarrow RO{:}CH_2 - CH_2\cdot$$

$$RO{:}CH_2 - CH_2\cdot + CH_2 = CH_2 \longrightarrow RO{:}CH_2 - CH_2 - CH_2 - CH_2\cdot$$

The reaction continues and gives

$$RO{:}CH_2CH_2 - (CH_2CH_2)_n CH_2CH_2\cdot$$

The reaction may end by one of two termination steps. One is the bonding of two free radicals

$$RO{:}CH_2CH_2 - (CH_2CH_2)_n CH_2CH_2\cdot$$
$$+ \cdot CH_2CH_2 - (CH_2CH_2)_n CH_2CH_2{:}OR$$
$$\downarrow$$
$$RO{:}CH_2CH_2 - (CH_2CH_2)_n CH_2CH_2{:}CH_2CH_2 - (CH_2CH_2)_n CH_2CH_2{:}OR$$

and the other is the internal stabilization of the polymer by double-bond formation.

$$RO{:}CH_2CH_2 - (CH_2CH_2)_n CH_2CH_2\cdot \longrightarrow$$
$$RO{:}CH_2CH_2 - (CH_2CH_2)_n CH = CH_2 + H\cdot$$

Molecular and Structural Formulas

The **alkynes** comprise a series of carbon- and hydrogen-based compounds that contain at least one triple bond. This group of compounds is a homologous series with the general molecular formula of C_nH_{2n-2}, where n equals any integer greater than one.

The simplest alkyne, **ethyne** (also known as acetylene), has two carbon atoms and the molecular formula of C_2H_2. The structural formula for ethyne is

$$H-C\equiv C-H$$

In longer alkyne chains, the additional carbon atoms are attached to each other by single covalent bonds. Each carbon atom is also attached to sufficient hydrogen atoms to produce a total of four single covalent bonds about itself. In alkynes of four or more carbon atoms, the triple bond can be located in different positions along the chain, leading to the formation of structural isomers. For example, the alkyne of molecular formula C_4H_6 has two isomers,

$$HC\equiv C-CH_2CH_3 \quad \text{and} \quad CH_3-C\equiv C-CH_3$$

Although alkynes possess restricted rotation due to the triple bond, they do not have stereoisomers like the alkenes because the bonding in a carbon-carbon triple bond is *sp* hybridized. In *sp* hybridization, the maximum separation between the hybridized orbitals is 180°, so the molecule is linear. Thus, the substituents on triple-bonded carbons are positioned in a straight line, and stereoisomers are impossible.

Unsaturation

The alkyne triple bond is composed of one σ and two π covalent bonds. The π bonds are the structures that preclude any rotation around the σ-bond axis. As with the alkenes, any rotation on the σ-bond axis would require disruption of the π-bond system. Breaking the π-bond system requires energy and thus would lead to a molecule with a less stable, higher energy state.

Like alkenes, alkynes are **unsaturated** because they are capable of reacting with hydrogen in the presence of a catalyst to form a corresponding fully saturated alkane. Each π bond signals that two hydrogen atoms have been lost from the molecular formula of the alkane with the same number of carbon atoms. Because alkynes possess two π bonds per molecule, they are said to contain two **elements of unsaturation.**

Nomenclature

Although some common alkyne names, such as acetylene, are still found in many textbooks, the International Union of Pure and Applied Chemistry (IUPAC) nomenclature is required for journal articles. The rules for alkynes in this system are identical with those for alkenes, except for the ending. The following rules summarize alkyne nomenclature.

1. Identify the longest continuous chain of carbon atoms that contains the carbon-carbon triple bond. The parent name of the alkyne comes from the IUPAC name for the alkane of the same number of carbon atoms, except the *-ane* ending is changed to *-yne* to signify the presence of a triple bond. Thus, if the longest continuous chain of carbon atoms containing a triple bond has five atoms, the compound is **pentyne.**

2. Number the carbon atoms of the longest continuous chain, starting at the end closest to the triple bond. Thus,

$$CH_3-CH_2-C\equiv C-CH_3$$

is numbered from right to left, placing the triple bond between the second and third carbon atoms of the chain. (Numbering the chain from left to right incorrectly places the triple bond between the third and fourth carbons of the chain.)

3. The position of the triple bond is indicated by placing the lower of the pair of numbers assigned to the triple-bonded carbon atoms in front of the name of the alkyne. Thus the compound shown in rule 2 is 2-pentyne.

4. The location and name of any substituent atom or group is indicated. For example, the compound

$$CH_3-\overset{\overset{\displaystyle H}{|}}{\underset{\underset{\displaystyle Cl}{|}}{C}}-CH_2-C\equiv C-CH_3$$

is 5-chloro-2-hexyne.

Physical Properties

The physical properties of alkynes are very similar to those of the alkenes. Alkynes are generally nonpolar molecules with little solubility in polar solvents, such as water. Solubility in nonpolar solvents, such as ether and acetone, is extensive. Like the alkanes and alkenes, alkynes of four or fewer carbon atoms tend to be gases.

Substituted alkynes have small dipole moments due to differences in electronegativity between the triple-bonded carbon atoms, which are sp hybridized, and the single-bonded carbon atoms, which are sp^3 hybridized. The sp-hybridized carbon atom, which possesses more s character than the sp^3-hybridized carbon atom, is more electronegative in character. The resulting asymmetrical electron distribution in the bond between such carbon atoms results in the generation of a dipole moment.

Acidity

Alkynes of the general structure

$$R-C\equiv C-H$$

are referred to as **terminal alkynes.** These types of alkynes are weakly acidic. Exposure to a strong base, such as sodium amide, produces an acid-base reaction.

$$\underset{\text{alkyne}}{R-C\equiv C-H} + \underset{\substack{\text{amide} \\ \text{ion}}}{NH_2^-} \rightleftharpoons \underset{\substack{\text{acetylide} \\ \text{ion}}}{R-C\equiv C^-} + \underset{\text{ammonia}}{NH_3}$$

The acidity of a terminal alkyne is due to the high level of s character in the sp hybrid orbital, which bonds with the s orbital of the hydrogen atom to form a single covalent bond. The high level of s character in an sp-hybridized carbon causes the overlap region of the σ bond to shift much closer to the carbon atom. This polarizes the bond, causing the hydrogen atom to become slightly positive. This slight positive charge makes the hydrogen atom a weak proton, which can be removed by a strong base.

In the case of alkanes and alkenes, the s character in the hybridized carbon bonds is less, resulting in fewer electronegative carbon atoms and a corresponding lesser shift toward those atoms in the overlap region of the σ bond. The location of the overlap region makes the corresponding hydrogen atoms less electron deficient and thus less acidic. In reality, the hydrogen atoms bonded to alkanes and alkenes can be removed as protons, but much stronger nonaqueous bases are necessary.

The reaction that forms the acetylide ion is reversible. Thus, the base may not form an acid of greater strength than the starting alkyne by acceptance of the proton, or the newly formed conjugate acid will reprotonate the acetylide ion. The fact that stronger acids are capable of reprotonating the acetylide ion can be seen in its reaction with water.

$$R-C\equiv C^- + H_2O \longrightarrow R-C\equiv C-H + OH^-$$

Preparations

The preparations of alkynes are very similar to those of the alkenes. The main preparative reactions involve the elimination of groups or ions from molecules, resulting in the formation of π bonds.

Dehydrohalogenation. The loss of a hydrogen atom and a halogen atom from adjacent alkane carbon atoms leads to the formation of an alkene. The loss of additional hydrogen and halogen atoms from the double-bonded carbon atoms leads to alkyne formation. The halogen atoms may be located on the same carbon (a **geminal dihalide**) or on adjacent carbons (a **vicinal dihalide**).

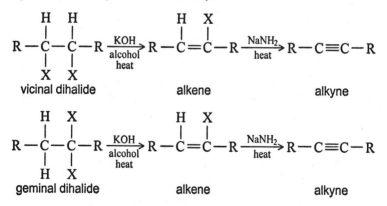

During the second dehydrohalogenation step, certain conditions are necessary, namely high temperatures and an extremely strong basic solution.

Dehalogenation. Vicinal tetrahaloalkanes can be dehalogenated with zinc metal in an organometallic reaction to form alkynes.

$$R-\underset{\underset{X}{|}}{\overset{\overset{X}{|}}{C}}-\underset{\underset{X}{|}}{\overset{\overset{X}{|}}{C}}-R + Zn \longrightarrow R-C\equiv C-R + ZnX_2$$

Substitution. Larger alkynes can be generated by reacting an alkyl halide with an acetylide ion, which is generated from a shorter alkyne.

$$R-C\equiv C^- + R'X \longrightarrow R-C\equiv C-R'$$

acetylide ion   alkyl halide

Because acetylide ions are bases, elimination reactions can occur, leading to the formation of an alkene from the alkyl halide. Because substitution and elimination reactions proceed through the formation of a common intermediate, these two types of reactions always occur simultaneously.

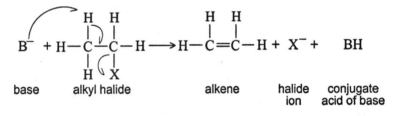

base   alkyl halide   alkene   halide ion   conjugate acid of base

Ethyne (acetylene) preparation. Ethyne, which is commonly called **acetylene,** is the simplest alkyne. Historically, it was prepared by reacting calcium carbide with water.

$$CaC_2 + H_2O \longrightarrow Ca(OH)_2 + HC\equiv CH$$

Today, ethyne is normally prepared by the pyrolysis of methane. In this procedure, a stream of methane gas is briefly heated to 1500°C in an airless chamber. Air must be excluded from the reaction or oxidation (combustion) will occur.

$$CH_4 \xrightarrow[0.1 \text{ s}]{1500°C} H-C\equiv C-H + H_2$$

Addition Reactions

The principal reaction of the alkynes is addition across the triple bond to form alkanes. These addition reactions are analogous to those of the alkenes.

Hydrogenation. Alkynes undergo catalytic hydrogenation with the same catalysts used in alkene hydrogenation: platinum, palladium, nickel, and rhodium. Hydrogenation proceeds in a stepwise fashion, forming an alkene first, which undergoes further hydrogenation to an alkane.

$$R-C\equiv C-R' \xrightarrow{H_2}{Pt} R-CH=CH-R' \xrightarrow{H_2}{Pt} R-CH_2-CH_2-R'$$

This reaction proceeds so smoothly that it is difficult, if not impossible, to stop the reaction at the alkene stage, although by using palladium or nickel for the catalyst, the reaction can be used to isolate some alkenes. Greater yields of alkenes are possible with the use of **poisoned catalysts.** One such catalyst, the **Lindlar catalyst,** is composed of finely divided palladium coated with quinoline and absorbed on calcium carbonate. This treatment makes the palladium less receptive to hydrogen, so fewer hydrogen atoms are available to react. When a catalyst is deactivated in such a manner, it is referred to as being **poisoned.**

The mechanism of alkyne hydrogenation is identical to that of the alkenes. Because the hydrogen is absorbed on the catalyst surface, it is supplied to the triple bond in a *syn* manner.

Alkynes can also be hydrogenated with sodium in liquid ammonia at low temperatures. This reaction is a chemical reduction rather than a catalytic reaction, so the hydrogen atoms are not attached to a surface, and they may approach an alkene from different directions, leading to the formation of *trans* alkenes.

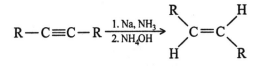

Halogenation. The addition of halogens to an alkyne proceeds in the same manner as halogen addition to alkenes. The halogen atoms add to an alkyne molecule in a stepwise fashion, leading to the formation of the corresponding alkene, which undergoes further reaction to a tetrahaloalkane.

$$R-C\equiv C-R + X_2 \xrightarrow{CCl_4} R-\underset{\displaystyle X}{\overset{\displaystyle X}{C}}=\underset{\displaystyle}{\overset{\displaystyle X}{C}}-R$$

$$R-\underset{\displaystyle X}{\overset{\displaystyle X}{C}}=\underset{\displaystyle X}{\overset{\displaystyle X}{C}}-R + X_2 \xrightarrow{CCl_4} R-\underset{\displaystyle X}{\overset{\displaystyle X}{C}}-\underset{\displaystyle X}{\overset{\displaystyle X}{C}}-R$$

Unlike most hydrogenation reactions, it is possible to stop this reaction at the alkene stage by running it at temperatures slightly below 0°C.

Hydrohalogenation. Hydrogen halides react with alkynes in the same manner as they do with alkenes.

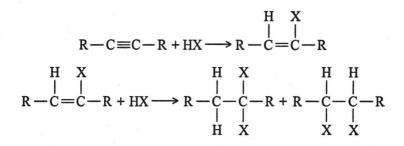

Both steps in the above addition follow the Markovnikov rule. Thus, the addition of hydrogen bromide to 1-butyne gives 2-bromo-1-butene as the major product of the first step.

CH_3— CH_2—$C≡CH$ + HBr ⟶
   1-butyne

$$CH_3—CH_2—\overset{\overset{\displaystyle Br}{|}}{C}=CH_2 + CH_3CH_2CH=CH_2Br$$
   2-bromo-1-butene   1-bromo-1-butene
   (major product)   (minor product)

The reaction of 2-bromo-1-butene in the second step gives 2,2-dibromobutane as the major product.

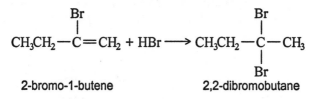

2-bromo-1-butene   2,2-dibromobutane

Hydration. The addition of the elements of water across the triple bond of an alkyne leads to the formation of aldehydes and ketones. Water addition to terminal alkynes leads to the generation of **aldehydes,** while nonterminal alkynes and water generate **ketones.**

These products are produced by rearrangement of an unstable **enol** (vinyl alcohol) intermediate. The term "enol" comes from the *en* in "alkene" and *ol* in "alcohol," reflecting that one of the carbon atoms in vinyl alcohol has both a double bond (alkene) and an OH group (alcohol) attached to it. A vinyl group is

$$CH_2{=}CH{-}$$

and a vinyl alcohol is

$$CH_2{=}CH{-}OH$$

Water adds across the triple bond of an alkyne via a carbocation mechanism. Dilute mineral acid and mercury(II) ions are needed for the reaction to occur.

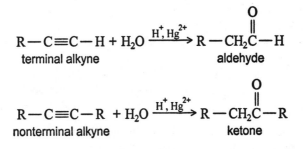

The first step of the mechanism is an acid-base reaction between the mercury(II) ion (Hg^{2+}) and the π system of the alkyne to form a π complex.

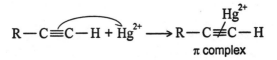

The π complex is converted into a single bond between one or the other of the carbons of the triple bond and the mercury(II) ion, with the resulting generation of a carbocation.

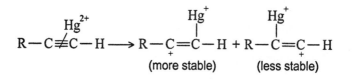

A molecule of water is attracted to the carbocation to form an oxonium ion.

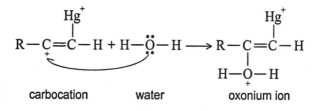

The oxonium ion loses a proton to stabilize itself.

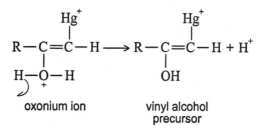

The vinyl alcohol precursor that results is converted into vinyl alcohol (enol) by reaction with a hydronium ion (H_3O^+).

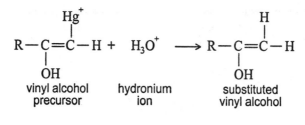

Vinyl alcohols (enols) are unstable intermediates, and they undergo rapid isomerization to form ketones. Such isomerization is called **keto-enol tautomerism.**

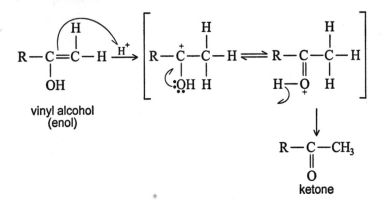

In a similar fashion, the less-stable intermediate generates an aldehyde.

Oxidation. Alkynes are oxidized by the same reagents that oxidize alkenes. Disubstituted alkynes react with potassium permanganate to yield vicinal diketones (vic-diketones or 1,2-diketones) or, under more vigorous conditions, carboxylic acids.

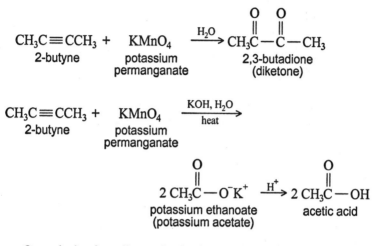

Ozonolysis of an alkyne also leads to carboxylic acid formation.

$$CH_3-CH_2-C\equiv C-CH_3 \xrightarrow[2.\ H^+]{1.\ O_3}$$
2-pentyne

$$\underset{\substack{\text{ethanoic acid}\\ \text{(acetic acid)}}}{CH_3-\overset{\displaystyle O}{\overset{\|}{C}}-OH} + \underset{\substack{\text{propanoic acid}\\ \text{(propionic acid)}}}{CH_3-CH_2-\overset{\displaystyle O}{\overset{\|}{C}}-OH}$$

Polymerization. Alkynes can be polymerized by both cationic and free-radical methods. The reactions and mechanisms are identical with those of the alkenes (see page 68).

Structural Formulas

Cyclohydrocarbons are the cyclic, or ring, analogues of the alkanes, alkenes, and alkynes. These hydrocarbons are often referred to as **alicyclic compounds,** and the simplest class is made up of the **cycloalkanes.** Their general molecular formula is C_nH_{2n}, where n equals any whole number of 3 or greater. Normally, these compounds are represented by geometric figures.

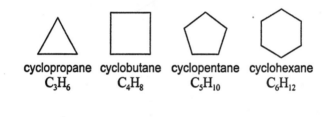

cyclopropane   cyclobutane   cyclopentane   cyclohexane
C_3H_6 C_4H_8 C_5H_{10} C_6H_{12}

Nomenclature

Substituted cycloalkanes are named in a manner similar to the open-chain, or aliphatic, alkanes. The following rules summarize the International Union of Pure and Applied Chemistry (IUPAC) nomenclature for substituted cycloalkanes.

1. Determine the number of carbon atoms in the ring and in the largest substituent. If the ring has more carbons than the substituent, the compound is an alkyl-substituted cycloalkane. If the substituent possesses more carbons than the ring, the compound is a cycloalkyl alkane.

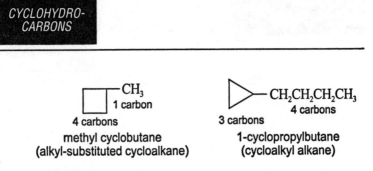

| methyl cyclobutane | 1-cyclopropylbutane |
| (alkyl-substituted cycloalkane) | (cycloalkyl alkane) |

2. If an alkyl-substituted cycloalkane has more than one substituent, the ring is numbered so the substituents have the lowest sum of numbers.

1,3-dimethylcyclohexane
not
1,5-dimethylcyclohexane

3. If the molecule possesses two or more different substituent groups, the number one position is determined by alphabetical priority.

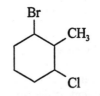

1-bromo-2-methyl-3-chlorohexane
not
1-chloro-2-methyl-3-bromohexane

Stereochemistry

Because ring compounds have restricted rotation around their carbon-carbon single bonds, cycloalkanes with two or more substituents have *cis* and *trans* stereoisomers. For example, 1,2-dibromobutane can exist as

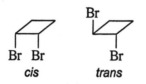

Cycloalkane rings of five or fewer carbon atoms form planar or very nearly planar rings. For such molecules, *cis* and *trans* isomerism refers to the location of substituents as either on the same side of the ring plane (*cis*) or on opposite sides of the ring plane (*trans*). Cycloalkanes that contain six or more carbons bend out of plane, or "pucker." For example, cyclohexane exists in three forms.

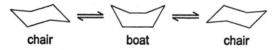

The two end structures are called **chair forms,** while the middle structure is called the **boat form.** The chair form of cyclohexane is more stable than the boat form.

Because the carbon atoms of cyclohexane form single bonds with each other, the bonds must be sp^3 hybridized. This type of hybridization creates a tetrahedral shape at each atom. Thus, the two bonds projecting from each carbon atom in the ring must form an angle of approximately 110°. This arrangement leads to the following structure.

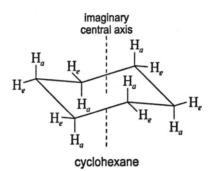

cyclohexane

The hydrogen atoms designated as H_a occupy **axial positions.** Their bonds roughly parallel an imaginary axis through the center of the carbon ring. The hydrogen atoms designated as H_e occupy **equatorial positions.** Each of these bonds is roughly 110° away from the axial bond on that carbon. *Cis* and *trans* substituent positions can be defined as those positions either on the same side (*cis*) or opposite sides (*trans*) of a plane that bisects the molecule through a designated pair of carbon atoms. For example, in 1,2-disubstitutions (the substituents are represented in the diagram below as A and B), the *cis*-1,2-isomer has one substituent in the axial position and one substituent in the equatorial position.

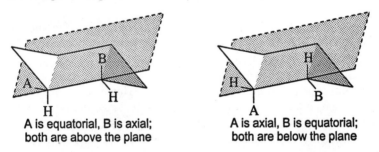

A is equatorial, B is axial;   A is axial, B is equatorial;
both are above the plane   both are below the plane

In both cases, the A and B groups are located on the same side of the plane. Thus, equatorial-axial substitutions lead to the *cis* isomer.

The *trans* isomer has both substituents in either axial or equatorial positions.

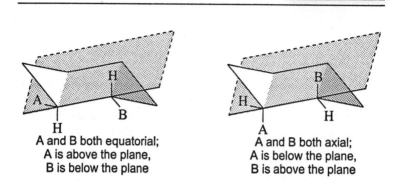

A and B both equatorial;
A is above the plane,
B is below the plane

A and B both axial;
A is below the plane,
B is above the plane

In both cases, the A and B groups are on opposite sides of the plane bisecting the adjacent carbons, creating *trans* isomers.

In 1,3-disubstitution, axial-axial or equatorial-equatorial arrangement generates *cis*-isomers, while axial-equatorial substitutions lead to *trans* isomers.

Finally, in 1,4-disubstitutions, axial-axial or equatorial-equatorial substitutions lead to the *trans*-isomer, while axial-equatorial substitutions generate the *cis*-isomer.

Reactivity and Stresses of Small Rings

Again, all cycloalkane ring carbon atoms are sp^3 hybridized, requiring bond angles that must be tetrahedral, or approximately 110°. However, three- and four-membered carbon rings are planar, so their bonding angles are 60° and 90°, respectively. The small size of these bond angles compared to the tetrahedral angle means that the orbital overlap region cannot exist directly between two carbon atoms. Rather, the two carbons are located at a slight angle to the overlap region, an arrangement that creates a weaker, more reactive bond. This type of bonding strain is called **angle strain.** Five-membered rings have a bond angle of 108°, which is very close to the tetrahedral angle. As a result, this ring system possesses little angle strain. Rings of six carbons or more bend and thus maintain the stable tetrahedral bonding angle.

In both the chair and boat forms of cyclohexane, there is no angle strain; however, the boat form has another type of ring strain called torsional strain. **Torsional strain** is caused by the interaction of hydrogen atoms or substituents that are bonded to either adjacent or nonadjacent carbon atoms and situated in an **eclipsed** fashion. The boat form of cyclohexane has two forms of torsional strain. The first type is caused by the interaction of atoms or groups that are eclipsed on adjacent carbons. It occurs between the four lower hydrogen atoms on the four carbons at the bottom of the boat form of cyclohexane. The second type is caused by eclipsed atoms or groups on nonadjacent carbons. This occurs between the eclipsed hydrogen atoms of the two upper carbons of the boat form. These two types of torsional strain account for the higher energy states of the eclipsed cycloalkanes compared to the cycloalkanes with staggered arrangements. Because the chair form of cyclohexane does not have torsional strain, it is more stable and has a lower energy state than the boat form.

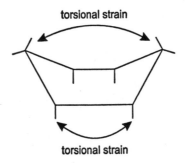

torsional strain

torsional strain

Preparations

Cycloalkanes can be prepared by ring-cyclization reactions, such as a modified Wurtz reaction or a condensation reaction. Additionally, they can be prepared from cycloalkenes and cycloalkynes (Figure 13).

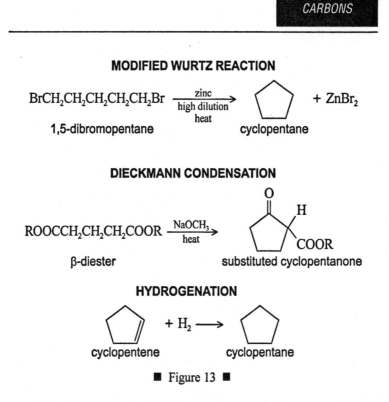

MODIFIED WURTZ REACTION

$$BrCH_2CH_2CH_2CH_2CH_2Br \xrightarrow[\text{heat}]{\substack{\text{zinc} \\ \text{high dilution}}} \bigcirc + ZnBr_2$$

1,5-dibromopentane   cyclopentane

DIECKMANN CONDENSATION

$$ROOCCH_2CH_2CH_2COOR \xrightarrow[\text{heat}]{NaOCH_3}$$

β-diester   substituted cyclopentanone

HYDROGENATION

$$\bigcirc + H_2 \longrightarrow \bigcirc$$

cyclopentene   cyclopentane

■ Figure 13 ■

Cycloalkenes and cycloalkynes are normally prepared from cycloalkanes by ordinary alkene-forming reactions, such as dehydration, dehalogenation, and dehydrohalogenation. Typical preparations for cyclohexene and cyclohexyne are illustrated in Figure 14.

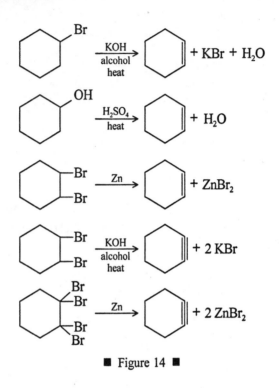

■ Figure 14 ■

Reactions

Due to angle strain, the bonds in three- and four-membered carbon rings are weak. Because of these weak bonds, cyclopropane and cyclobutane undergo reactions that are atypical of alkanes. For example, cyclopropane reacts with halogens dissolved in carbon tetrachloride to form dihaloalkanes.

$$\triangle + Br_2 \xrightarrow{CCl_4} BrCH_2CH_2CH_2Br$$

Under similar conditions, straight-chain propane does not react.

In general, cycloalkanes undergo the normal reactions of the **aliphatic alkanes** (the straight-chain and branched-chain alkanes). Thus, cyclopentane will react with halogens in ultraviolet light to form halosubstituted cycloalkanes.

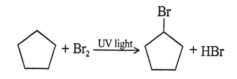

$$CH_3CH_2CH_3 + Br_2 \xrightarrow{\text{UV light}} CH_3CH_2CH_2Br + HBr$$

Cycloalkenes and cycloalkynes undergo the ordinary addition reactions of alkenes and alkynes. Cyclopropene, cyclopropyne, cyclobutene, and cyclobutyne also undergo ring-opening reactions.

Stereochemistry is the study of the three-dimensional structure of molecules. The previously mentioned *cis* and *trans* isomers are forms of **stereoisomers,** differing structurally only in the location of the atoms of the molecule in three-dimensional space. Such stereoisomers can have different physical and chemical properties. For example, the *cis* and *trans* isomers of butenedioic acid show vast differences in their physical and chemical properties.

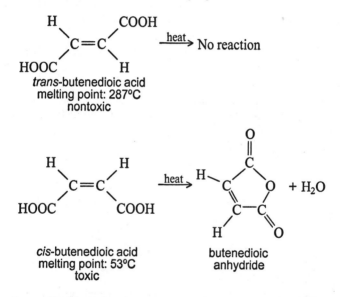

Stereochemistry is of particular interest to biochemists because the reactivity and toxicity of molecules change with their stereochemistry. Most body reactions are **stereospecific,** meaning that receptor sites on cells accept only molecules with specific spatial arrangements of their atoms. Other configurations of the same chemical either will not react or may be toxic to the living being.

There are other types of stereoisomers in addition to the *cis* and *trans* arrangements of atoms. To understand these other isomers, you must first understand the three-dimensional structures of the molecules.

Optical Activity

The stereochemistry of molecules has its roots in the work of the nineteenth-century French physicist Jean Biot. Biot was studying the nature of plane-polarized light (Figure 15) when he discovered that solutions of some organic compounds caused polarized light to rotate. Compounds with this property are called **optically active,** and the amount and direction of rotation can be determined with a polarimeter (Figure 16).

White light consists of electromagnetic waves oscillating in an infinite number of planes at right angles to the direction of light travel. **Plane-polarized light** is ordinary white light that has passed through a polarizer. A **polarizer** is a filter that blocks the light waves in all planes but one. A common example is a pair of Polaroid sunglasses, which prevent glare by polarizing sunlight. Figure 15 illustrates the concept of plane-polarized light.

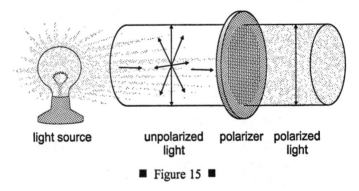

light source   unpolarized   polarizer   polarized
                     light   light

■ Figure 15 ■

Figure 16 shows how the optical activity of a compound is measured in a polarimeter.

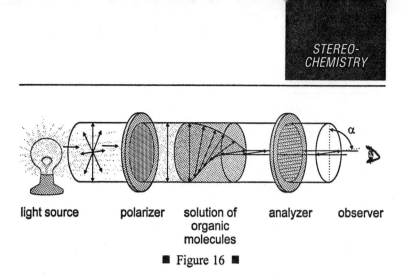

light source   polarizer   solution of   analyzer   observer
organic
molecules

■ Figure 16 ■

In 1849, while doing recrystallization experiments with tartaric acid salts, Louis Pasteur isolated two distinct kinds of crystals. Upon microscopic examination, he discovered the crystals were nonsuperimposable mirror images of each other. Pasteur was able to separate the different crystals into two piles. Solutions of the pure samples of each kind of crystal showed optical activity. The angle of rotation (α) in a polarimeter was the same for both solutions, but the directions of rotation were opposite. A solution of the original mixture of crystals showed no optical activity. The crystalline structure of the tartrate salt crystals is illustrated in Figure 17.

SODIUM AMMONIUM TARTRATE

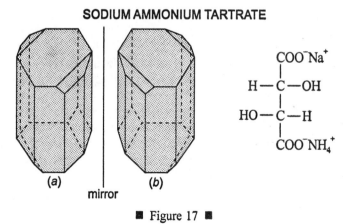

(a)   (b)
mirror

■ Figure 17 ■

As you can see, the crystals are nonsuperimposable, having the same relationship as a right hand to a left hand. Such opposite configurations are called **enantiomers;** thus, the two forms of tartrate crystals are enantiomers of each other.

Optical purity. A collection of molecules of one enantiomer is said to be **optically pure,** while a 1:1 mixture of two enantiomers is a **racemic mixture.** Between these two extremes, there can be an infinite number of mixtures containing various ratios of two enantiomers. Such chemical mixtures have a specified optical purity. The **optical purity** of a mixture of enantiomers is defined as the ratio of the rotation of the mixture to the rotation of the pure enantiomers.

$$\text{optical purity} = \frac{\text{observed rotation of mixture}}{\text{rotation of pure enantiomer}} \times 100$$

Chirality

Molecules that form nonsuperimposable mirror images, and thus exist as enantiomers, are said to be **chiral molecules.** For a molecule to be chiral, it cannot contain a plane of symmetry. A **plane of symmetry** is a plane that bisects an object (a molecule, in this case) in such a way that the two halves are identical mirror images. An example of a structure that has a plane of symmetry is a cylinder. Cutting a cylinder in half lengthwise generates two halves that are exact mirror images of each other. A molecule that possesses a plane of symmetry in any of its conformations is identical to its own mirror image. Such molecules are **achiral,** or **nonchiral.** Butane is an achiral molecule, while 2-bromobutane is chiral.

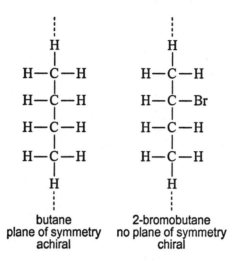

butane
plane of symmetry
achiral

2-bromobutane
no plane of symmetry
chiral

The most common cause of chirality in an organic molecule is a carbon atom with four different atoms or groups bonded to it. This carbon atom is called a **stereogenic, chiral,** or **asymmetric center.** (Such centers are often designated with an asterisk in formulas and projections.)

The **van't Hoff rule** predicts the maximum number of enantiomers an optically active molecule can possess. This rule states that the maximum number of enantiomers a molecule can have is equal to 2 raised to the nth power, where n equals the number of stereogenic centers. The molecule 2-chlorobutane has one stereogenic center, so two enantiomers are possible.

$$CH_3 - \overset{\overset{\textstyle Cl}{|}}{\underset{\underset{\textstyle H}{|}}{C^*}} - CH_2CH_3 \qquad CH_3CH_2 - \overset{\overset{\textstyle Cl}{|}}{\underset{\underset{\textstyle H}{|}}{C^*}} - CH_3$$

mirror
*stereogenic center
Maximum number of enantiomers = $2^n = 2^1 = 2$

Projections

An explanation of the various drawings, or **projections,** used to show the three-dimensional structure of chemicals will help you understand the next section's discussion of enantiomers and diastereomers. The simplest drawing is called a **Fischer projection.** In this representation, the "backbone" atoms of a carbon chain are represented simply by a straight line, and the terminal carbons are written as groups. The atoms or groups bonded to the chain carbons are "attached" with perpendicular lines. Compare the structural formula and Fischer projection of 2-bromo-3-chlorobutane.

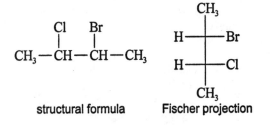

structural formula   Fischer projection

A second type of projection, a **sawhorse projection,** allows better visualization of the three-dimensional geometry between adjacent carbon atoms. This projection is customarily used to show interactions between groups on adjacent carbon atoms in mechanisms. In a sawhorse projection, the backbone carbons are represented by a diagonal line, and the terminal carbons are shown in groups, just as in the Fischer projection. You can see in the next illustration that the top carbon group of the Fischer projection of 2-bromo-3-chlorobutane has become the back carbon group of the sawhorse projection.

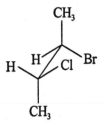

A sawhorse projection can reveal staggered and eclipsed conformations in molecules. The previous conformation of 2-bromo-3-chlorobutane is referred to as a **staggered conformer** because the atoms and groups attached to each backbone carbon fit in the voids around the groups on the adjacent carbon. In an **eclipsed conformer,** the groups and atoms on adjacent carbons are in line with each other.

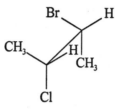

A third type of projection is called the **Newman projection.** This type of projection is used mainly to show interaction leading to stress between atoms or groups in three-dimensional space due to steric crowding. In this representation, a molecule is viewed from one "end." The "front" carbon of the backbone is represented by a dot, and the "back" carbon of the backbone is shown as a circle. Lines representing the bonds attaching the atoms and groups to the backbone carbons emanate from the dot and the circle at 120° angles. Compare the two conformations of 2-bromo-3-chlorobutane in the Newman projection to the previous sawhorse and Fischer projections.

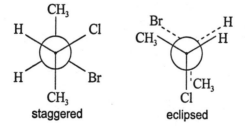

staggered   eclipsed

Enantiomers and Diastereomers

A Fischer projection is the most useful projection for discovering enantiomers. Compare the 2-chlorobutane enantiomer structures in this diagram.

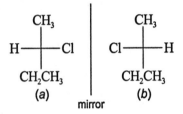

Rotating structure (*b*) 180° in the plane of the paper, the only allowable rotation, does not lead to a form that is superimposable on structure (*a*). Rotations of less than or more than 180° are not allowed because in a two-dimensional projection, it is impossible to see the difference in the position of atoms that are located in front of or behind the plane.

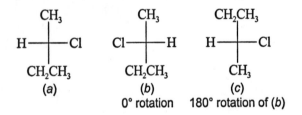

Structures (*a*) and (*b*) are the only pair of enantiomers for 2-chlorobutane.

The compound 2-chloro-3-bromobutane has two stereogenic centers and a maximum of four enantiomers. Compare these two Fischer projections.

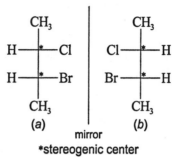

*stereogenic center

Structure (b) cannot be superimposed on structure (a) by rotating it in the plane of the page, so structures (a) and (b) are enantiomers. The additional two enantiomers are created by allowing rotation about one of the stereogenic centers while restricting rotation about the other. Structure (c) is created by allowing rotation about the upper stereogenic center (carbon 2) of structure (a).

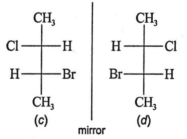

Notice that structure (c) has a different configuration from structures (a) and (b). Structure (d), the mirror image of (c), cannot be superimposed on structure (c) by rotating it in the plane of the page. Therefore, structures (c) and (d) are enantiomers. Any further rotation about the stereogenic centers creates a structure that is already drawn. For example, starting with structure (a) and allowing rotation about the lower stereogenic center (carbon 3) generates structure (d) again. This situation agrees with the maximum number of enantiomers predicted by the van't Hoff rule: $2^n = 2^2 = 4$.

The relationship between the enantiomers of separate enantio-morphic pairs is called **diastereoisomerism.** For example, while structures (*a*) and (*b*), and (*c*) and (*d*), are enantiomers, the relationship of (*a*) to (*c*) is one of diastereoisomerism. They are not mirror images, so structure (*a*) is a **diastereomer** of structures (*c*) and (*d*). Likewise, structure (*b*) is a diastereomer of structures (*c*) and (*d*). In the same fashion, structures (*c*) and (*d*) are diastereomers of (*a*) and (*b*). **Enantiomers** have opposite configurations at all stereogenic centers, while **diastereomers** have the same configuration at one or more stereogenic centers but opposite configurations at others.

Optically inactive stereogenic centers (*meso* forms). Some molecules are optically inactive even though they contain stereogenic centers. These compounds normally contain a **plane of symmetry.** The compound 2,3-dichlorobutane should have four enantiomers because it has two stereogenic centers.

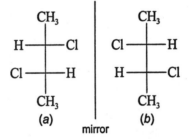

Structure (*b*) cannot be superimposed on structure (*a*) by rotating it in the plane of the page; thus, structures (*a*) and (*b*) are enantiomers. Rotation about the upper stereogenic center leads to structure (*c*), which is a different configuration from (*a*) and (*b*).

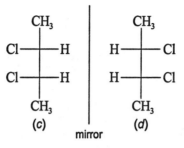

Structure (d) is the mirror image of (c). It can be superimposed on (c) by rotating it 180°. Because these two structures are superimposable mirror images, they are not optically active, even though they contain two stereogenic centers. The reason for this lack of optical activity is the plane of symmetry through the center of the molecule.

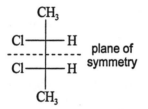

These types of molecules are called *meso* forms. In **meso forms,** the stereogenic centers are optically active, but due to the molecular symmetry, they rotate plane-polarized light to the same degree but in opposite directions. This phenomenon results in an internal cancellation of optical activity.

Racemic Mixtures and the Resolution of Enantiomers

Enantiomorphic pairs show no rotation of plane-polarized light if they are in a true 1:1 mixture. Again, such mixtures are referred to as **racemic mixtures,** or **racemates.**

Racemic mixtures can be separated, or **resolved,** into their pure enantiomers by three methods. The first method is to mechanically separate the crystals in such a mixture based on differences in their shapes. This was the method first used by Pasteur, and it is mainly of historical interest.

The second resolution method employs enzymes. **Enzymes** are stereospecific chiral protein molecules that act as catalysts. Because of their chirality, these molecules react with only one enantiomer in a racemic mixture. The enantiomer that momentarily bonds to an enzyme undergoes reaction, while the enantiomer that does not bond remains unchanged. The unreacted enantiomer can then be removed from the reaction mix by ordinary separation methods, such as distillation or recrystallization.

The third method involves converting the enantiomers of a racemic mixture into diastereomers and then resolving that mixture with ordinary separation techniques. The separated diastereomers are then treated with appropriate reagents to regenerate the original enantiomers.

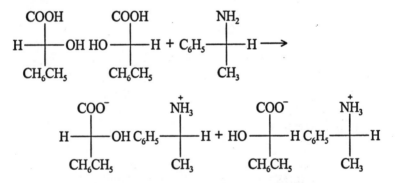

In this example, the diastereomer salts are separated by recrystallization, and the original acids are regenerated by the addition of a hydrochloric acid solution.

Cahn-Ingold-Prelog *RS* Notational System

Because enantiomers are different configurations of the same compound, a notational system had to be developed that would indicate the three-dimensional arrangement of atoms at specific stereogenic centers. Such a system was devised by the chemists Cahn, Ingold, and Prelog. In this system, the substituents of a stereogenic center are ranked by atomic weight as dictated by a series of **priority rules.** A projection of the molecule is then viewed so that the group or atom of lowest priority is eclipsed by the stereogenic center. The ranking of the three remaining groups is then determined. If their rank from highest to lowest is in a clockwise direction, the configuration is *R*. On the other hand, if the rank declines in a counterclockwise direction, the configuration is *S*. The labels *R* and *S* come from the Latin words *rectus,* which means "right," and *sinister,* meaning "left." The right and left designations refer only to the order of atoms or groups about a stereogenic center. They do not refer to the direction in which plane-polarized light is rotated by the molecule.

The direction of rotation of plane-polarized light by a molecule is designated "d" or "+" for **dextrorotatory compounds,** which rotate plane-polarized light to the right, and "l" or "−" for **levorotatory compounds,** which rotate plane-polarized light to the left.

The following **sequence rules** summarize the Cahn-Ingold-Prelog notational system.

1. Identify the four different atoms or groups attached to the stereogenic center.

2. Rank the atoms or groups based on the priority rules (see the list following this one).

3. Orient a projection of the molecule in space so that the group or atom of lowest rank is eclipsed by the stereogenic center.

4. Determine the ranking of the remaining visible atoms or groups.

5. If the ranking declines in a clockwise direction, the configuration is *R;* if the ranking declines in a counterclockwise direction, the configuration is *S.*

The **priority rules** rank atoms and groups based on atomic mass. The following list summarizes these rules.

1. For the four atoms directly attached to the stereogenic center, the higher the atomic mass, the higher the rank.

2. If two or more atoms directly attached to the stereogenic center have the same mass, work outward along the chains of the groups they are in, atom by atom, until a point of difference is reached. The rank is assigned at this point of difference, based on the difference in atomic mass.

3. If a group contains multiple bonds, the doubly or triply bonded atoms are counted as two or three of those atoms, respectively. Thus the carbonyl group

$$\diagup C = O$$

is considered to have two carbon-oxygen bonds, one actual and one theoretical.

A cyano group

$$-C \equiv N$$

is considered to have three carbon-nitrogen bonds, one actual and two theoretical. For comparison purposes, an actual bond ranks higher than a theoretical bond of the same type. For example, when ranking the cyano group against

the displayed group takes priority, due to its three actual carbon-nitrogen bonds.

The sequence rules can be illustrated by applying them to lactic acid.

1. The four atoms or groups around the stereogenic carbon are CH_3, H, COOH, and OH.

2. The ranking based on atomic weights is oxygen > carbon > hydrogen. The ranking of the carbon-containing groups is COOH > CH_3.

3. The overall priority is thus OH > COOH > CH_3 > H. Viewing the molecule in such a way that the lowest ranked atom is eclipsed by the stereogenic center reveals that the remaining groups decline in rank in a clockwise direction. Thus, the stereogenic carbon has an *R* configuration.

Gaining the proper perspective for eclipsing the lowest ranked atom or group by the stereogenic carbon can be difficult. Try using an eclipsed downward sawhorse projection of the molecule. This projection allows rotation of the sawhorse axis so the atom or group of lowest rank can be placed at the bottom, thus allowing the point of view to be always from the top, or front, of the projection. Such a sawhorse projection must be drawn directly from a Fischer projection without any rotation. An eclipsed downward sawhorse projection of (*R*)-lactic acid can be drawn this way.

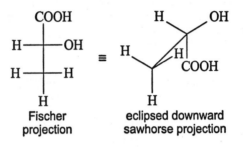

Fischer
projection

eclipsed downward
sawhorse projection

The projection can be abbreviated in this manner.

Rotation on the carbon-carbon bond to put the hydrogen atom at the bottom of the projection looks like this.

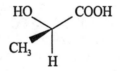

Viewing from the top eclipses the hydrogen atom.

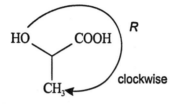

The clockwise direction of the ranking is then easily seen.

RS notation can also be generated from Fischer projections; however, because Fischer projections are strictly two-dimensional representations of three-dimensional molecules, only certain manipulations are allowed. One such manipulation is the interchange of any two pairs of substituents. An interchange converts an enantiomer to its mirror image. Interchanging two pairs of substituents produces the original projection. Here is an example using a Fischer projection of (*R*)-lactic acid.

Verify its *R* or *S* configuration by using the following general steps (diagrams follow).

1. Interchange the atom or group of lowest rank with the atom or group at the bottom of the Fischer projection.

2. Interchange the atoms or groups on the remaining positions until rank order is established.

3. Determine whether the ranking defines a clockwise or counterclockwise direction. If clockwise, the *projection* is an *R* configuration; if counterclockwise, it is an *S* configuration.

4. The configuration of the *original structure* is determined by counting the number of interchanges made in step 2. If an odd number were made, the original configuration is opposite that of the projection. If an even number of interchanges were made, the original configuration and the projection are the same.

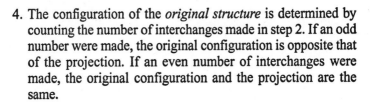

The ranking of this projection of (R)-lactic acid is S, and because it was arrived at after an odd number of switches (one), the original configuration is really R. The enantiomer of (R)-lactic acid is (S)-lactic acid, which has the configuration

Try applying the four general steps to this configuration of lactic acid.

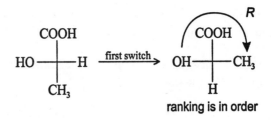

The final projection is R. However, this configuration was found after an odd number of exchanges (one); thus, the original configuration must be S.

Stereochemistry of Reactions

Many simple reactions involve stereochemistry. This section briefly reviews and summarizes the stereochemistry of reactions already discussed.

A reaction that involves only achiral reactants and reagents can produce only racemic mixtures or products that are achiral. Thus, the monobromination of butane produces a mixture of 1-bromobutane and 2-bromobutane. The 1-bromobutane has no stereogenic center and is, therefore, achiral. The 2-bromobutane, which has a stereogenic center, is a mixture of the enantiomers (R)-2-bromobutane and (S)-2-bromobutane. This racemic mixture results because the reaction proceeds via a free-radical mechanism. Free radicals are almost perfectly planar. This planarity makes the product mixture achiral because there is equal probability of a bromine free radical attacking from either side of the carbon free radical intermediate (see the chapter "Reactions of Alkanes").

Addition of a halogen atom to an alkene proceeds via a *trans* addition. This occurs due to steric hindrance created by the formation of the halonium ion (see the chapter "Reactions of Alkenes").

Catalytic hydrogenation of alkenes and alkynes takes place via *cis* addition because absorption of the hydrogen on the surface of the rigid catalyst allows the hydrogen atoms to approach the double bond from only one side of the alkene molecule (again, see the chapter "Reactions of Alkenes").

Dehydrohalogenation of chloroethane to form ethene proceeds via a *trans* elimination reaction because of the tetrahedral shape of the carbon atom bonded to the chlorine atom. To displace the chlorine as a chloride ion, the electron pair of the other carbon must approach from the unhindered side opposite the chlorine atom.

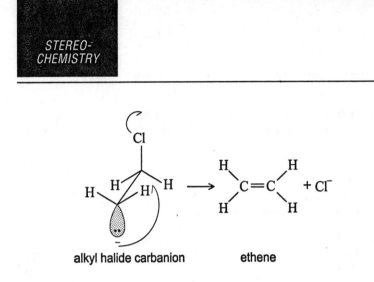

alkyl halide carbanion   ethene

Notice the use of the sawhorse projection to show how this reaction occurs.

A **diene** is a molecule that has two double bonds. If the molecule is also a hydrocarbon, it is called an **alkadiene.** When the double bonds are separated from each other by two or more single bonds, they are called **isolated double bonds.** Isolated double bonds undergo normal alkene reactions, revealing that no interaction occurs between them. If, however, the double bonds are separated by only one single bond, atypical reactions occur. Such an arrangement is called a **conjugated double-bond system.** The interaction between the two double bonds in conjugated dienes delocalizes the electron density and increases the stability of the molecule. The simplest conjugated diene is 1,3-butadiene.

$$H_2C=CH-CH=CH_2$$

Stability of Conjugated Systems

Investigations of 1,3-butadiene have shown that the central single bond is slightly shorter than expected. In addition, the heat of hydrogenation of the molecule, 57.1 kilocalories per mole, is less than the amount predicted from doubling the heat of hydrogenation of two butene molecules (60.6 kcal/mol).

A molecular orbital picture of 1,3-butadiene shows possible π-bond overlap throughout the molecule.

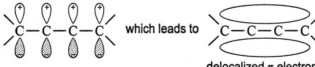

delocalized π electrons

For this type of delocalized bonding to occur, the atomic *p* orbitals must align, with all the lobes having the same phase signs. Alignment of oppositely signed lobes leads to a higher energy state.

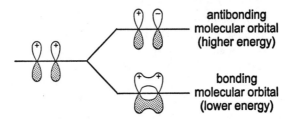

Thus, a conjugated diene system must look like this.

This arrangement produces a low-energy state. The complete delocalization of the π system gives the single bond some double-bond character and explains why it is slightly shorter than expected. Rotation around this single bond is also somewhat restricted because of the partial double-bond character of the bond and because of an increase in repulsion between groups attached to the terminal carbons. The increase in repulsive forces is due to the shorter bond length, which brings the groups closer together.

Preparations

Dienes are prepared from the same reactions that form ordinary alkenes. The two most common methods are the dehydration of diols (dihydroxy alkanes) and the dehydrohalogenation of dihalides (dihaloalkanes). The generation of either an isolated or conjugated system depends on the structure of the original reactants. Vicinal diols, which have two hydroxyl groups on adjacent carbon atoms, and vicinal dihalides, which have halogen substituents on adjacent car-

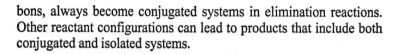

bons, always become conjugated systems in elimination reactions. Other reactant configurations can lead to products that include both conjugated and isolated systems.

Dehydration of diols:

$$HOCH_2CH_2CH_2CH_2OH \xrightarrow{H^+} H_2C=CH-CH=CH_2$$
butane-1,4-diol   1,3-butadiene

$$HOCH_2CH_2CH_2CH_2CH_2OH \xrightarrow{H^+} H_2C=CH-CH_2-CH=CH_2$$
pentane-1,5-diol   1,4-pentadiene

Dehydrohalogenation of dihalides:

$$XCH_2CH_2CH_2CH_2X \xrightarrow[\text{alcohol}]{KOH} H_2C=CH-CH=CH_2$$
1,4-dihalobutane   1,3-butadiene

$$XCH_2CH_2CH_2CH_2CH_2X \xrightarrow[\text{alcohol}]{KOH} H_2C=CH-CH_2-CH=CH_2$$
1,5-dihalopentane   1,4-pentadiene

1,2 and 1,4 Additions

Both isolated and conjugated dienes undergo electrophilic addition reactions. In the case of isolated dienes, the reaction proceeds in a manner identical to alkene electrophilic addition. The addition of hydrogen bromide to 1,4-pentadiene leads to two products.

$$H_2C=CHCH_2CH=CH_2 + HBr \longrightarrow$$

$$\overset{\displaystyle Br}{\underset{\displaystyle |}{H_3CCHCH_2CH=CH_2}} + BrH_2CCH_2CH_2CH=CH_2$$
4-bromo-1-pentene   5-bromo-1-pentene
(Markovnikov product)   (anti-Markovnikov product)

This reaction follows the standard carbocation mechanism for addition across a double bond. The addition of more hydrogen bromide results in addition across the second double bond in the molecule. In the case of conjugated dienes, a 1,4-addition product forms in addition to the Markovnikov and anti-Markovnikov products. Thus, in the addition of hydrogen bromide to 1,3-butadiene, the following occurs.

$$H_2C{=}CH{-}CH{=}CH_2 \xrightarrow{\text{HBr}} CH_3{-}\overset{\displaystyle\overset{\text{Br}}{|}}{CH}{-}CH{=}CH_2$$

1,3-butadiene   3-bromo-1-butene
(Markovnikov product)

$$+\,Br{-}CH_2{-}CH_2{-}CH{=}CH_2 + CH_3{-}CH{=}CH{-}CH_2{-}Br$$

4-bromo-1-butene   1-bromo-2-butene
(anti-Markovnikov product)   (1,4-addition product)

The 1,4-addition product is the result of the formation of a stable allylic carbocation. An **allylic carbocation** has the structure

$$CH_2{=}CH{-}\overset{+}{C}H_2$$

It is very stable because the charge on the primary carbon is delocalized along the carbon chain by movement of the π electrons in the π bond. This delocalization of charge by electron movement is called **resonance,** and the various intermediate structures are called **resonance structures.** However, according to resonance theory, none of the intermediate resonance structures are correct. The true structure is a hybrid of all the structures that can be drawn. The **hybrid structure** contains less energy and is thus more stable than any of the resonance structures. The more resonance structures that can be drawn for a given molecule, the more stable it is. The difference in energy between the calculated energy content of a resonance structure and the actual energy content of the hybrid structure is called the **resonance energy, conjugation energy,** or **delocalization energy** of the molecule. The allylic carbocation exists as a hybrid of two resonance structures.

$$CH_2\!=\!CH\!-\!\overset{+}{C}H_2 \longleftrightarrow \overset{+}{C}H_2\!-\!CH\!=\!CH_2$$

Because it is resonance stabilized, the allylic carbocation is much more stable than an ordinary primary carbocation. Resonance stability always leads to a more stable state than inductive stability. The hybrid structure for this ion is

$$[CH_2\!=\!=\!CH\!=\!=\!CH_2]^+$$

This structure shows the π-electron movement throughout the conjugated system, with a resulting delocation of the positive charge through the system.

Understanding the allylic carbocation clarifies the mechanism for the addition to 1,3-butadiene.

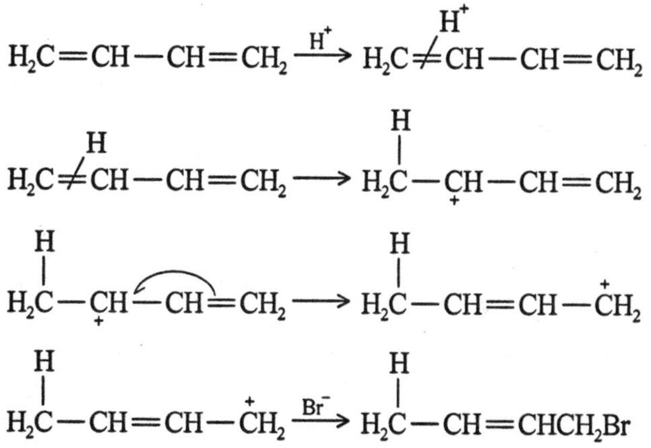

$$H_2C\!=\!CH\!-\!CH\!=\!CH_2 \xrightarrow{\ H^+\ } H_2C\!\overset{H^+}{\underset{}{\neq}}CH\!-\!CH\!=\!CH_2$$

$$H_2C\overset{H}{\underset{}{\neq}}CH\!-\!CH\!=\!CH_2 \longrightarrow H_2\overset{H}{\underset{}{C}}\!-\!\underset{+}{C}H\!-\!CH\!=\!CH_2$$

$$H_2\overset{H}{\underset{}{C}}\!-\!\underset{+}{C}H\!-\!CH\!=\!CH_2 \longrightarrow H_2\overset{H}{\underset{}{C}}\!-\!CH\!=\!CH\!-\!\overset{+}{C}H_2$$

$$H_2\overset{H}{\underset{}{C}}\!-\!CH\!=\!CH\!-\!\overset{+}{C}H_2 \xrightarrow{\ Br^-\ } H_2\overset{H}{\underset{}{C}}\!-\!CH\!=\!CHCH_2Br$$

When other electrophiles are added to conjugated dienes, 1,4 addition also occurs. Many reactants, such as halogens, halogen acids, and water, can form 1,4-addition products with conjugated dienes. Whether more 1,2 addition or 1,4 addition product is created depends largely on the temperature at which the reaction is run. For example, the addition of hydrogen bromide to 1,3-butadiene at temperatures below zero leads mainly to the 1,2-addition product, while addition reactions run at temperatures above 50°C with these chemicals produces mainly the 1,4-addition product. If the reaction is initially run at 0°C and then warmed to 50°C or higher and held there for a period of time, the major product will be a 1,4 addition. These results indicate that the reaction proceeds along two distinct pathways. At high temperatures, the reaction is thermodynamically controlled, while at low temperatures, the reaction is kinetically controlled.

For the general reaction

$$R \longrightarrow A + B$$

the high-temperature, thermodynamically controlled reaction exists in an equilibrium state.

$$A \rightleftharpoons R \rightleftharpoons B$$

If B is more stable than A, B will be the major product formed. The rate of formation is immaterial because an increase in the forward reaction rate is mirrored by an increase in the reverse reaction rate. In reversible reactions, the product depends only on thermodynamic stability.

At low temperatures, the reaction is irreversible and no equilibrium is established because the products have insufficient energy to overcome the activation energy barrier, which separates them from the initial reactant. If A forms faster than B, it will be the major product. In irreversible reactions, the product depends only on the reaction rate and is therefore said to be kinetically controlled. Figure 18 is a reaction energy diagram that illustrates thermodynamically and kinetically controlled reactions.

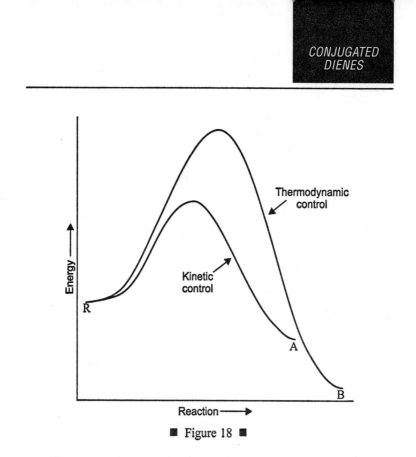

■ Figure 18 ■

The energy diagram of the reaction of 1,3-butadiene with hydro-gen bromide shows the pathways of the two products generated from the intermediate (Figure 19).

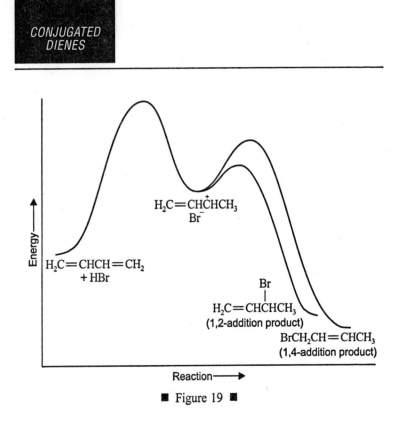

Figure 19

Diels-Alder Reaction

The **Diels-Alder reaction** is a cycloaddition reaction between a conjugated diene and an alkene. This reaction produces a 1,4-addition product. A typical example is the reaction of 1,3-butadiene with maleic anhydride.

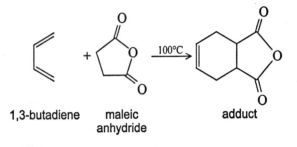

The Diels-Alder reaction is favored by the presence of electron-withdrawing groups on the diene and electron-releasing groups on the dienophile, which is a group or bond that is attracted to a diene. The mechanism for the Diels-Alder reaction shows that it does not run via a carbocation intermediate. Instead, this reaction proceeds by a **pericyclic process,** a mechanism of just one step, involving a cyclic redistribution of bonding electrons.

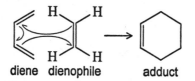

diene   dienophile   adduct

Simple alkenes and alkynes are not good dienophiles. A good dienophile generally has one or more electron-withdrawing carbonyl groups

$$\text{C}=\text{O}$$

or cyano groups

$$-\text{C}\equiv\text{N}$$

attached to it. Electron-supplying groups on the diene make the electrons of the π system more available for reaction.

Diels-Alder stereochemistry. The Diels-Alder reaction is very stereospecific. The original stereochemistry of the diene and the dienophile are preserved during this *syn*-addition reaction. An example of this stereospecificity is the reaction of 1,3-butadiene with *cis*-diethylmaleate.

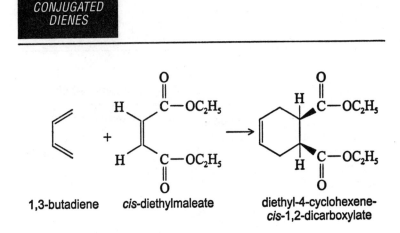

| 1,3-butadiene | cis-diethylmaleate | diethyl-4-cyclohexene-cis-1,2-dicarboxylate |

Because the reaction is basically a concerted cyclization, the diene must react in the *cis* conformation.

If the diene is a ring structure, the Diels-Alder reaction produces a bicyclic ring system.

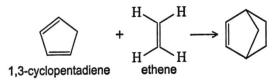

1,3-cyclopentadiene   ethene

A **bicyclic ring system** has two carbon rings that share common sides. The previous diagram shows what appears to be a cyclohexene ring with a carbon bridge connecting the third and sixth carbons. In reality, this system is two five-membered rings, a cyclopentene ring and a cyclopentane ring, which are sharing two sides (the "carbon bridge"). The structure of this molecule is

Alkanes

Hydrogenation of alkenes

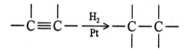

$$-\overset{|}{C}=\overset{|}{C}-\xrightarrow{\underset{Pt}{H_2}}-\overset{|}{\underset{|}{C}}-\overset{|}{\underset{|}{C}}-$$

Hydrogenation of alkynes

$$-C\equiv C-\xrightarrow{\underset{Pt}{H_2}}-\overset{|}{\underset{|}{C}}-\overset{|}{\underset{|}{C}}-$$

Reduction of alkyl halides

$$R-X\xrightarrow{\underset{Zn}{HX}}R-H$$

Coupling of alkyl halides (Wurtz reaction)

$$2R-X\xrightarrow{Na}R-R$$

$$2R-X\xrightarrow{\underset{2.\ CuI}{1.\ Li}}R-R$$

Hydrolysis of Grignard reagent

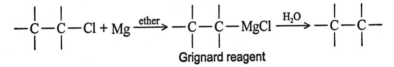

Grignard reagent

Alkenes

Dehydrohalogenation of alkyl halides

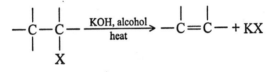

Dehydration of alcohols

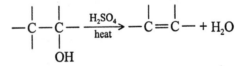

Dehalogenation of vicinal dihalides

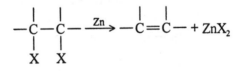

Alkynes

Dehydrohalogenation of vicinal and geminal dihalides

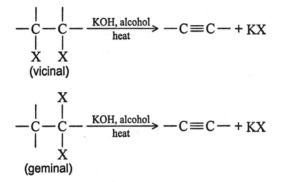

Dehalogenation of vicinal tetrahaloalkanes

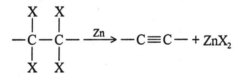

Substitution

$$R-C\equiv C-H \xrightarrow{NH_2^-} R-C\equiv C^- \xrightarrow{R'X} R-C\equiv C-R'$$

Cycloalkanes

Carbene insertion

$$-\overset{|}{C}=\overset{|}{C}- \xrightarrow[\text{heat}]{\text{CH}_2\text{N}_2} \triangle + \text{N}_2$$

Simmons-Smith reaction

$$-\overset{|}{C}=\overset{|}{C}- \xrightarrow[\text{Zn(Cu)}]{\text{CH}_2\text{X}_2} \triangle + \text{ZnX}_2$$

Modified Wurtz reaction

$$X-\overset{|}{\underset{|}{C}}-\overset{|}{\underset{|}{C}}-\overset{|}{\underset{|}{C}}-\overset{|}{\underset{|}{C}}-\overset{|}{\underset{|}{C}}-X \xrightarrow[\text{high dilution}]{\text{Zn}} \pentagon$$

Dieckmann condensation

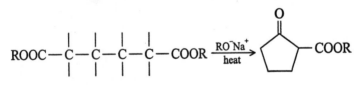

$$\text{ROOC}-\overset{|}{\underset{|}{C}}-\overset{|}{\underset{|}{C}}-\overset{|}{\underset{|}{C}}-\overset{|}{\underset{|}{C}}-\text{COOR} \xrightarrow[\text{heat}]{\text{RO}^-\text{Na}^+} \text{—COOR}$$

Alkanes

Oxidation

$$-\overset{|}{\underset{|}{C}}-\overset{|}{\underset{|}{C}}- \xrightarrow{O_2} CO_2 + H_2O$$

Halogenation

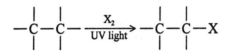

$$-\overset{|}{\underset{|}{C}}-\overset{|}{\underset{|}{C}}- \xrightarrow[\text{UV light}]{X_2} -\overset{|}{\underset{|}{C}}-\overset{|}{\underset{|}{C}}-X$$

Alkenes

Halogenation

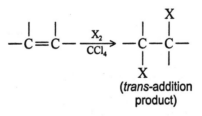

$$-\overset{|}{C}=\overset{|}{C}- \xrightarrow[\text{CCl}_4]{X_2} -\overset{|}{\underset{|}{C}}-\overset{\overset{\displaystyle X}{|}}{\underset{|}{C}}-$$
$$\underset{X}{}$$
(*trans*-addition
product)

Hydrohalogenation

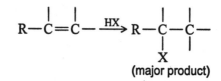

(major product)

(The reaction follows the Markovnikov rule.)

Addition of hydrogen bromide in the presence of peroxide

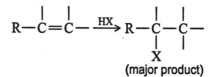

(major product)

(This is an anti-Markovnikov addition.)

Hydration (direct addition of water)

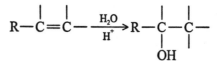

(The reaction follows the Markovnikov rule.)

Hydroboration-oxidation (indirect addition of water)

R—C=C— $\xrightarrow[\text{NaOH}]{\text{B}_2\text{H}_6,\,\text{H}_2\text{O}_2}$ R—C—C—
OH

Hydrogenation

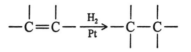

Epoxide formation

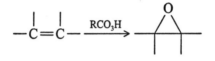

Oxidation

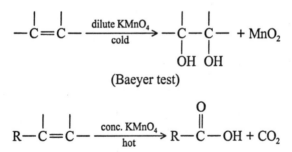

(Baeyer test)

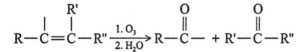

Ozonolysis

$$R-\overset{|}{C}=\overset{\overset{R'}{|}}{C}-R'' \xrightarrow[\text{2. H}_2\text{O}]{\text{1. O}_3} R-\overset{O}{\overset{||}{C}}- \ + \ R'-\overset{O}{\overset{||}{C}}-R''$$

Cationic polymerization

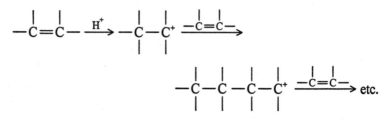

Free-radical polymerization

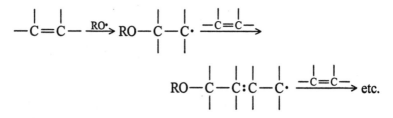

Addition of halogens and water (hypohalous acid)

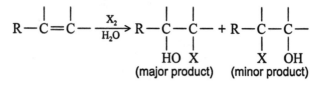

Alkynes

Hydrogenation

$$-C\equiv C- \xrightarrow[\text{Pt}]{\text{H}_2} -C=C- \xrightarrow[\text{Pt}]{\text{H}_2} -C-C-$$

Hydrohalogenation

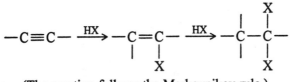

(The reaction follows the Markovnikov rule.)

Halogenation

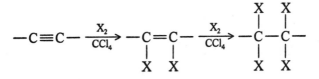

Hydration (keto-enol tautomerization)

$$R-C\equiv C- \xrightarrow[H^+]{H_2O,\ Hg^{2+}} R-\overset{\overset{\displaystyle O}{\|}}{C}-\overset{|}{\underset{|}{C}}-$$

Reaction of acidic terminal hydrogen (acid-base reaction)

$$R-C\equiv C- \xrightarrow{NH_2^-} R-C\equiv C^- + NH_3$$

Cycloalkanes

Opening reactions of three- and four-membered rings

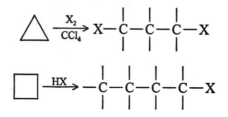

Halogenation

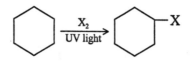

Oxidation

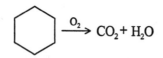

Glossary

achiral the opposite of *chiral;* also called *nonchiral.* An achiral molecule can be superimposed on its mirror image.

acid see **Arrhenius theory, Brønsted-Lowry theory of acids and bases,** and **Lewis theory of acids and bases.**

acid-base reaction a neutralization reaction in which the products are a salt and water.

activated complex molecules at an unstable intermediate stage in a reaction.

activation energy the energy that must be supplied to chemicals to initiate a reaction; the difference in potential energy between the ground state and the transition state of molecules. Molecules of reactants must have this amount of energy to proceed to the product state.

addition a reaction that produces a new compound by combining all of the elements of the original reactants.

adduct the product of an addition reaction.

alcohol an organic chemical that contains an — OH group.

aldehyde an organic chemical that contains a — CHO group. Water addition to terminal alkynes forms aldehydes.

alicyclic compound an *ali*phatic *cyclic* hydrocarbon, which means that an alicyclic compound contains a ring but not an aromatic benzene ring (a six-carbon ring with three double bonds).

aliphatic compound a straight- or branched-chain hydrocarbon; an alkane, alkene, or alkyne.

alkane a hydrocarbon that contains only single covalent bonds. The alkane general formula is C_nH_{2n+2}.

alkene a hydrocarbon that contains a carbon-carbon double bond. The alkene general formula is C_nH_{2n}.

alkoxide ion an anion formed by removing a proton from an alcohol; the RO^- ion.

alkoxy free radical a free radical formed by the homolytic cleavage of an alcohol — OH bond; the RO· free radical.

alkyl group an alkane molecule from which a hydrogen atom has been removed. Alkyl groups are abbreviated as "R" in structural formulas.

alkyl halide a hydrocarbon that contains a halogen substituent, such as fluorine, chlorine, bromine, or iodine.

alkyl-substituted cycloalkane a cyclic hydrocarbon to which one or more alkyl groups are bonded. (Compare with "cycloalkyl alkane.")

alkyne a hydrocarbon that contains a triple bond. The alkyne general formula is C_nH_{2n-2}.

allyl group the $H_2C = CHCH_2 —$ group.

allylic carbocation the $H_2C = CHCH_2^+$ ion.

analogue in organic chemistry, chemicals that are similar to each other, but not identical. For example, the hydrocarbons are all similar to each other, but an alkane is different from the alkenes and alkynes because of the types of bonds they contain. Therefore, an alkane and an alkene are analogues.

angle of rotation (α) in a polarimeter, the angle right or left in which plane-polarized light is turned after passing through an optically active compound in solution.

angle strain the strain created by the deformation of bond angles from their normal values.

angular momentum quantum number (l) the second number in Schrödinger's electron wave equation, which tells the shape of the orbital.

anion a negatively charged ion.

***anti* addition** the addition of atoms to opposite sides of a molecule. (Compare with "*syn* addition.")

antibonding molecular orbital a molecular orbital that contains more energy than the atomic orbitals from which it was formed; in other words, an electron is less stable in an antibonding orbital than it is in its original atomic orbital.

anti-Markovnikov addition a reaction in which the hydrogen atom of a hydrogen bromide bonds to the carbon of a double bond that is bonded to *fewer* hydrogen atoms. The addition takes place via a free-radical intermediate rather than a carbocation. (Compare with "Markovnikov rule.")

Arrhenius theory a theory (limited to aqueous systems) that defines an acid as a compound that liberates hydrogen ions and a base as a compound that liberates hydroxide ions. A *neutralization* is the reaction of a hydrogen ion with a hydroxide ion to form water.

asymmetric center see **stereogenic center.**

atom the smallest amount of an element; a nucleus surrounded by electrons.

atomic mass (A) the sum of the weights of the protons and neutrons in an atom. (A proton and neutron each have a mass of 1 atomic mass unit.)

atomic number (Z) the number of protons or electrons in an atom.

atomic 1s orbital the spherical orbital nearest the nucleus of an atom.

atomic orbital a region in space around the nucleus of an atom where the probability of finding an electron is high.

atomic p orbital an hourglass-shaped orbital, oriented on x, y, and z axes in three-dimensional space.

atomic s orbital a spherical orbital.

aufbau buildup the order in which electrons fill atomic orbitals according to energy factors.

axial bond a bond positioned perpendicularly to the general plane of a cyclohexane ring.

axial position the position a group occupies in an axial bond. (See **axial bond.**)

Baeyer reagent cold, dilute potassium permanganate, which is used to oxidize alkenes and alkynes.

base see **Arrhenius theory, Brønsted-Lowry theory of acids and bases,** and **Lewis theory of acids and bases.**

β elimination a group of reactions that form double or triple bonds through the loss of atoms or groups from adjacent carbon atoms; included are dehydrations, dehydrogenations, dehydrohalogenations, and double dehydrohalogenations.

bicyclic ring system a molecule made up of two carbon rings that share common sides.

boat conformation one of the conformations of the cyclohexanes, which resembles a boat. The boat conformation has high energy because of interactions between eclipsed hydrogen atoms or groups.

bond angle the angle formed between two adjacent bonds on the same atom.

bond-dissociation energy the amount of energy needed to homolytically fracture a bond.

bonding electron see **valence electron.**

bonding molecular orbital the orbital formed by the overlap of adjacent atomic orbitals.

bond length the equilibrium distance between the nuclei of two atoms or groups that are bonded to each other.

bond strength the amount of energy needed to homolytically fracture a bond (also called **bond-dissociation energy**).

branched-chain alkane an alkane with alkyl groups bonded to the central carbon chain.

Brønsted-Lowry theory of acids and bases a Brønsted-Lowry acid is a compound capable of donating a proton (a hydrogen ion), and a Brønsted-Lowry base is capable of accepting a hydrogen ion. In a *neutralization,* an acid donates a proton to a base, creating a conjugate acid and a conjugate base.

Cahn-Ingold-Prelog notational system a system that labels the three-dimensional arrangement of atoms around a stereogenic center of a molecule by ranking their atomic weights. The letter *R* indicates a clockwise decline in rank, and *S* indicates a counterclockwise decline in rank.

carbanion a carbon atom bearing a negative charge; a carbon anion.

carbene an electrically uncharged molecule that contains a carbon atom with only two single bonds and just six electrons in its valence shell.

carbenoid a chemical that resembles a carbene in its chemical reactions.

carbocation a carbon cation; a carbon atom bearing a positive charge (sometimes referred to as a "carbonium ion").

carbonyl group the $\diagdown C = O$ group.

carboxylic acid the $-\underset{\underset{O}{\|}}{C}-OH$ group.

catalyst a substance that affects the rate of a reaction in which it participates; however, it is not altered or used up in the process. Platinum metal is a catalyst in alkyne hydrogenation, for example.

cation a positively charged ion.

cationic polymerization see **polymerization;** cationic polymerization occurs via a cation intermediate and is less efficient than free-radical polymerization.

chain reaction a reaction that, once started, produces sufficient energy to keep the reaction running. These reactions proceed by a series of steps, which produce intermediates, energy, and products.

chair conformation a conformation of cyclohexane that resembles a chair and has less energy than a boat conformation.

chiral describes a molecule that is not superimposable on its mirror image; like the relationship of a left hand to a right hand.

chiral center see **stereogenic center.**

chiral molecule a molecule that has a chiral center and rotates plane-polarized light.

cis **isomer** a stereoisomer in which substituents are located on the same side of a double bond. (Compare with "*trans* isomer.")

competing reactions two reactions that start with the same reactants but form different products.

concerted taking place at the same time without the formation of an intermediate.

condensation reaction a reaction in which two molecules join to form a product and release a small molecule such as water.

condensed formula a formula in which the single bonds between the atoms are not shown with lines.

configuration the specific arrangement of atoms and groups in three-dimensional space. A configuration is characteristic of a specific stereoisomer.

conformation a specific three-dimensional shape of a molecule at any given time. Conformation can change by rotation around a single bond.

conjugate acid the acid that results when a Brønsted-Lowry base accepts a hydrogen ion.

conjugate base the base that results when a Brønsted-Lowry acid loses a hydrogen ion.

conjugated double bonds carbon-carbon double bonds that are separated from one another by just one single bond; for example, C=C—C=C.

conjugation the overlapping in all directions of a series of p orbitals. This process usually occurs in a molecule with alternating double and single bonds.

conjugation energy see **resonance energy.**

constitutional isomer see **structural isomer.**

covalent bond a bond formed by the sharing of electrons between atoms.

cyano group the — C≡N group.

cyclization the formation of ring structures.

cycloaddition a reaction that forms a ring.

cycloalkane a ring hydrocarbon made up of carbon and hydrogen atoms joined by single bonds.

cycloalkyl alkane an alkane to which a cyclical structure is bonded.

cyclohydrocarbon an alkane, alkene, or alkyne formed in a ring structure rather than a straight or branched chain.

debye unit (D) the unit of measure for a dipole moment. One debye equals 1.0×10^{-18} esu · cm. (See **dipole moment.**)

dehalogenation the elimination reaction in which two halogen atoms are removed from adjacent carbon atoms to form a double bond.

dehydration the elimination reaction in which a molecule of water is removed from a molecule.

dehydrohalogenation the elimination reaction in which a hydrogen atom and a halogen atom are removed from a molecule to form a double bond.

delocalization the spreading of electron density or electrostatic charge across a molecule.

delocalization energy see **resonance energy.**

deprotonation the loss of a proton (hydrogen ion) from a molecule.

dextrorotatory describes the clockwise rotation of plane-polarized light (from Latin *dextro*, "to the right"). A lowercase "*d*" or a

"+" is the notation used before an isomer's name to indicate that it is dextrorotatory; for example, *d*-2-butanol. (Compare with "levorotatory.")

diastereomer a stereoisomer of an optical isomer that has more than one stereogenic center and is not a mirror image of one of the other enantiomers of the molecule. Diastereomers have the same configuration at one or more stereogenic centers but opposite configurations at others.

Dieckmann condensation a condensation is a reaction in which two molecules join to form a new product, eliminating water or some other small molecule in the process. A Dieckmann condensation takes place within one molecule, resulting in a new cyclical molecule and the elimination of a small molecule.

Diels-Alder reaction a cycloaddition reaction between a conjugated diene and an alkene that produces a 1,4-addition product.

diene an organic compound that contains two double bonds.

dienophile the alkene that adds to the diene in a Diels-Alder reaction.

dihalide a compound that contains two halogen atoms; also called a "dihaloalkane."

diol a compound that contains two hydroxyl (— OH) groups; also called a "dihydroxy alkane."

dipole moment a measure of the polarity of a molecule; it is the mathematical product of the charge in electrostatic units (esu) and the distance that separates the two charges in centimeters (cm). For example, substituted alkynes have dipole moments caused by differences in electronegativity between the triple-bonded and single-bonded carbon atoms.

distillation the separation of components of a liquid mixture based on differences in boiling points.

double bond a multiple bond composed of one σ bond and one π bond. Rotation is not possible around a double bond. Hydrocarbons containing one double bond are called *alkenes,* and hydrocarbons with two double bonds are called *dienes.*

duet two electrons. Helium, the simplest noble gas, has a duet of electrons. The gaining of an electron by a hydrogen atom adds stability because it achieves the helium duet. (Compare with "octet.")

E **(entgegen)** the notation for the stereochemical arrangement in which the higher-ranked substituent groups are on opposite sides of the double bond. (See ***E-Z* notation.**)

eclipsed conformation one of the possible orientations of atoms or groups around two carbon atoms joined by a single bond. Atoms and groups bonded to an eclipsed conformer are positioned in line with one another, creating repulsive forces that give the molecule a high energy state. (Compare with "staggered conformation.")

electromagnetic wave the type of wave found in visible, infrared, and ultraviolet light as well as radio signals and X-rays.

electron negatively charged particles of little weight that exist in quantized probability areas around the atomic nucleus.

electron affinity the amount of energy liberated when an electron is added to an atom in the gaseous state.

electron dot structure a system in which the entire structure of the atom, except its valence electrons, is represented by the symbol for the element. The valence electrons are represented by dots (also known as a **Lewis structure**).

electronegativity the measure of an atom's ability to attract electrons toward itself in a covalent bond. The halogen fluorine is the most electronegative element.

electronegativity scale an arbitrary scale by which the electronegativity of individual atoms can be compared.

electrophile an "electron seeker;" an atom that seeks an electron to stabilize itself.

electrophilic addition a reaction in which the addition of an electrophile to an unsaturated molecule results in the formation of a saturated molecule.

electrostatic attraction the attraction of a positive ion for a negative ion.

element of unsaturation a π bond; a multiple bond or ring in a molecule.

enantiomer a stereoisomer that cannot be superimposed on its mirror image.

enantiomorphic pair in optically active molecules with more than one stereogenic center, the two structures that are mirror images of each other are enantiomorphic pairs.

energy of reaction the difference between the total energy content of the reactants and the total energy content of the products. The greater the energy of reaction, the more stable the products.

enol an unstable compound (for example, vinyl alcohol) in which a hydroxide group is attached to a carbon bearing a carbon-carbon double bond. These compounds tautomerize to form ketones, which are more stable.

enthalpy of activation see **activation energy.**

enzyme a stereospecific chiral protein that acts as a biological catalyst.

epoxide a three-membered ring that contains oxygen.

equatorial bond a bond attached to a ring structure that roughly parallels the equator of the ring.

equatorial position the position a group occupies in an equatorial bond.

equilibrium constant a measure of the degree of completion of an equilibrium reaction.

equivalent orbitals orbitals of the same principal level and type, such as the three p orbitals.

ester the
$$\overset{\text{O}}{\overset{\|}{-\text{C}}}-\text{OR}$$
functional group.

ether an organic compound in which an oxygen atom is bonded to carbon atoms. The general formula is R — O — R′. Epoxyethane, an epoxide, is a cyclic ether.

excited state a higher energy state than the ground state, achieved by adding energy to an atom or molecule in its ground state.

exothermic describes the giving off of energy as heat.

E-Z notation a notation, somewhat like the *cis* and *trans* system, that is used for alkenes with more than two substituents. The atoms or groups on either side of the double bond are ranked by atomic weight. If the heavier atoms are on the same side of the molecule, it is labeled *Z*, and if the heavier atoms are on opposite sides of the molecule, it is labeled *E*.

Fischer projection a projection that uses perpendicular lines to depict the absolute configuration of chiral molecules on a planar surface.

formal charge a charge on an atom created by the loss or gain of electrons.

four-center interaction a reaction in which bonds are simultaneously formed and broken between four atoms. For example, the reaction A — A + B — B could form two A — B molecules by simultaneously forming the new bonds while breaking the old bonds.

free radical an atom or group that has a single unshared electron.

free-radical chain reaction a reaction that proceeds by a free-radical intermediate in a chain mechanism, which is a series of self-propagating, interconnected steps. (Compare with "free-radical reaction.")

free-radical polymerization a polymerization initiated by a free radical.

free-radical reaction a reaction in which a covalent bond is formed by the union of two radicals. (Compare with "free-radical chain reaction.")

functional group a set of bonded atoms that displays a specific molecular structure and chemical reactivity when bonded to a carbon atom in the place of a hydrogen atom.

geminal a term that describes the location of two identical atoms or groups as being bonded to the same carbon atom; a geminal dihalide, for example. (Compare with "vicinal.")

glycol a class of alcohols that contain two — OH groups; $C_nH_{2n}(OH)_2$.

Grignard reagent an organometallic reagent in which magnesium metal inserts between an alkyl group and a halogen; for example, CH_3MgBr.

ground state the most stable electron configuration for an atom; this configuration has the least energy associated with it.

haloalkane an alkane that contains one or more halogen atoms; also called an alkyl halide.

halogen an electronegative, nonmetallic element in Group VII of the periodic table, including fluorine, chlorine, bromine, and iodine. Halogens are often represented in structural formulas with an "X."

halogenation a reaction in which halogen atoms are bonded to an alkene at the double bond.

halonium ion a halogen atom that bears a positive charge. This ion is highly unstable.

heat of combustion the energy released when an alkane is completely oxidized.

heat of hydrogenation the energy released when two hydrogen atoms bond to the carbon atoms of a former double bond.

hetero atom in organic chemistry, an atom other than carbon.

heterocyclic compound a class of cyclic compounds in which one of the ring atoms is not carbon; epoxyethane, for example.

heterogenic bond formation a type of bond formed by the overlap of orbitals on adjacent atoms. One orbital of the pair donates both of the electrons to the bond.

heterolytic cleavage the fracture of a bond in such a manner that one of the atoms receives both electrons. In reactions, this asymmetrical bond rupture generates carbocation and carbanion mechanisms.

homogenic bond formation a type of bond formation in which each atom donates one electron to the bond.

homologous series a set of compounds with common compositions; for example, the alkanes.

homologue one of a series of compounds in which each member differs from the next by a constant unit.

homolytic cleavage the fracture of a bond in such a manner that both of the atoms receive one of the bond's electrons. This symmetrical bond rupture forms free radicals; in reactions, it generates free-radical mechanisms.

hybrid atomic orbital a probability area created by a linear combination of atomic orbitals.

hybridization the changing, or mixing, of orbitals to form new atomic or molecular orbitals that are lower in energy.

hybrid orbital an orbital formed by the linear combination of atomic orbitals in the ground state.

hybrid orbital number rule the hybrid orbital number is equal to the sum of a molecule's σ bonds plus the number of unshared electron pairs. A hybrid orbital number of 2 indicates *sp*

---

hybridization; 3 indicates sp^2 hybridization; 4 indicates sp^3 hybridization.

hydration the addition of the elements of water to a molecule.

hydride shift the movement of a hydride ion, a hydrogen atom with a negative charge, to form a more inductively stabilized carbocation.

hydroboration-oxidation the addition of borane (BH_3) or an alkyl borane to an alkene and its subsequent oxidation to produce the anti-Markovnikov indirect addition of water.

hydrocarbon a molecule that contains exclusively carbon and hydrogen atoms. The central bond may be a single, double, or triple covalent bond, and it forms the backbone of the molecule.

hydrogenation the addition of hydrogen to an unsaturated compound.

hydrohalogenation a reaction in which a hydrogen atom and a halogen atom are added to a double bond to form a saturated compound.

hydrolyze to cleave a bond via the elements of water.

inductive effect the electron-donating or electron-withdrawing effect that is transmitted through σ bonds. It can also be defined as the ability of an alkyl group to "push" electrons away from itself. The inductive effect gives stability to carbocations and makes tertiary carbocations the most stable.

initiation step the first step in the mechanism of a reaction.

initiator a material capable of being easily fragmented into free radicals, which in turn initiate a free-radical reaction.

insertion placing between two atoms.

in situ from Latin, meaning "in place."

intermediate a species that forms in one step of a multistep mechanism; intermediates are unstable and cannot be isolated.

ion a charged atom; an atom that has either lost or gained electrons.

ionic bond a bond formed by the transfer of electrons between atoms, resulting in the formation of ions of opposite charge. The electrostatic attraction between these ions is the ionic bond.

ionization energy the energy needed to remove an electron from an atom.

isoelectronic having the same number of electrons. For example, a sodium atom that is lacking one electron is isoelectronic with neon, an inert gas.

isolated double bond a double bond that is more than one single bond away from another double bond in a diene.

isomer compounds that have the same molecular formula but different structural formulas.

IUPAC nomenclature a systematic method for naming molecules based on a series of rules developed by the International Union of Pure and Applied Chemistry. IUPAC nomenclature is not the only system in use, but it is the most common.

keto-enol tautomerization the process by which an enol equilibrates with its corresponding aldehyde or ketone.

ketone a compound in which an oxygen atom is bonded to a carbon atom, which is itself bonded to two more carbon atoms. Water addition to nonterminal alkynes forms ketones.

kinetically controlled reaction a reaction in which the rate of formation of the competing products accounts for the major product.

kinetic control reactions that have a major product that forms the fastest are under kinetic control. These reactions follow the lowest activation-energy pathway.

kinetics the study of reaction rates.

levorotatory describes the counterclockwise rotation of plane-polarized light (from Latin, *levo,* "to the left"). A lowercase "*l* " or "–" is the notation used before an isomer's name to indicate it is levorotatory; for example, *l*-2-butanol. (Compare with "dextrorotatory.")

Lewis structure see **electron dot structure.**

Lewis theory of acids and bases a Lewis acid is a compound capable of accepting an electron pair, and a Lewis base is capable of donating an electron pair.

Lindlar catalyst a particular poisoned catalyst used in alkyne reactions; it is finely divided palladium coated with quinoline and absorbed on barium sulfate.

linear the shape of a molecule with *sp* hybrid orbitals; an alkyne.

linear combination of atomic orbitals the process of combining atomic orbitals to form new orbitals. Linear combination can occur between orbitals in a single atom, creating hybrid atomic orbitals, or between the orbitals of two atoms, creating molecular orbitals. In either case, the number of orbitals always remains constant.

line-bond structure a representation of a molecule that shows covalent bonds as lines between atoms.

lone-pair electrons a nonbonding pair of electrons, which occupy the valence orbitals.

magnetic quantum number (*m*) the third number in Schrödinger's electron wave equation, which tells the orientation of the orbital in space.

major product the product that forms in the greatest amount in a reaction.

Markovnikov rule states that the positive part of a reagent (a hydrogen atom, for example) adds to the carbon of the double bond that already has more hydrogen atoms attached to it. The negative part adds to the other carbon of the double bond. Such an arrangement leads to the formation of the more stable carbocation over other less-stable intermediates.

mass number the total number of protons and neutrons in an atom.

mechanism the series of steps that reactants go through during their conversion into products.

***meso* compound** a compound that has a stereogenic center but is optically inactive because it also has a plane of symmetry.

methylene group a — CH_2 group.

minor product the product that forms in the least amount in a reaction.

molecular formula a chemical formula that shows the number and kinds of atoms in a molecule but not their arrangement; for example, C_2H_6.

molecular orbital an orbital formed by the linear combination of two atomic orbitals.

molecule a covalently bonded collection of atoms that has no electrostatic charge.

monomer the smallest molecule that reacts with itself to form a polymer.

multiple bond a double or triple bond; multiple bonds involve the atomic *p* orbitals in side-to-side overlap, preventing rotation.

neutralization the reaction of an acid and a base. The products are a salt and water.

neutron an uncharged particle in the atomic nucleus that has the same weight as a proton. Additional neutrons do not change an element but convert it to one of its isotopic forms.

Newman projection a drawing of a molecule that shows a head-on view of a carbon-carbon bond. The front carbon is represented by a dot, and the rear carbon is represented by a circle. Substituents are shown as spokes radiating from the dot or circle. This projection is used to show the possible interactions of substituents bonded to adjacent carbon atoms.

node a region of zero electron density in an orbital; a point of zero amplitude in a wave.

nonbonding electrons valence electrons that are not used for covalent bond formation.

nonchiral see **achiral.**

nonterminal alkyne an alkyne in which the triple bond is located somewhere other than the 1 position.

nucleophile a species that is capable of donating a pair of electrons to a nucleus.

nucleus the central core of an atom; the location of the protons and neutrons.

octet eight electrons. Carbon, oxygen, and the halogens either share, lose, or gain electrons to have eight electrons in their valence shells. (Compare with "duet.")

optical activity the ability of some chemicals to rotate plane-polarized light.

optical isomer another name for an **enantiomer.**

optical purity a number equal to the angle of rotation of a solution divided by the rotation of the pure enantiomer ($\times$ 100).

orbit an area around an atomic nucleus where there is a high probability of finding an electron; also called a shell. An orbit is divided into orbitals, or subshells.

orbital an area in an orbit where there is a high probability of finding an electron; a subshell. All of the orbitals in an orbit have the same principal and angular quantum numbers.

organometallic reaction a reaction in which a metallic element adds between a carbon atom and an electronegative atom in an organic molecule.

outer-shell electron see **valence electron.**

overlap region the region in space where atomic or molecular orbitals overlap, creating an area of high electron density.

oxidation the loss of electrons by an atom in a covalent bond. In organic reactions, this occurs when a compound accepts additional oxygen atoms.

oxirane a three-membered ring that contains oxygen; also called an epoxide ring.

oxonium ion a positively charged oxygen atom.

ozonide a compound formed by the addition of ozone to a double bond.

ozonolysis the cleavage of double and triple bonds by ozone, O_3.

paired spin the spinning in opposite directions of the two electrons in a bonding orbital.

parent name the root name of a molecule according to the IUPAC nomenclature rules; for example, hexane is the parent name in *trans*-1,2-dibromocyclohexane.

Pauli exclusion principle states that no two electrons in an atom can have the same set of quantum numbers.

pericyclic process a step in a reaction in which the bonding electrons are redistributed through a cyclic structure.

peroxide a compound that contains an oxygen-oxygen single covalent bond.

phase sign the positive and negative symbols assigned to the upward and downward displacement, respectively, of the standing wave that describes the orbitals about an atom's nucleus. Each upward and downward displacement is called a phase.

π (pi) bond a bond formed by the side-to-side overlap of atomic *p* orbitals. A π bond is weaker than a σ bond because of poor orbital overlap caused by nuclear repulsion. Unsaturated molecules are created by π bonds.

π complex an intermediate formed when a cation is attracted to the high electron density of a π bond.

π molecular orbital a molecular orbital created by the side-to-side overlap of atomic *p* orbitals.

plane of symmetry an imaginary plane that bisects a molecule, producing two halves that are mirror images of each other.

plane-polarized light ordinary light that has had all oscillations of the electromagnetic field filtered out but one. The remaining oscillation exists in only one plane.

poisoned catalyst a deactivated catalyst; one that is less effective in reactions than the nonpoisoned material.

polar covalent bond a bond in which the shared electrons are not equally available in the overlap region, leading to the formation of partially positive and partially negative ends on the molecule.

polarimeter a device that first polarizes light and then passes the polarized light through a chemical solution. An analyzer shows the degree and direction of rotation of the plane-polarized light if the chemical is optically active.

polarity the asymmetrical distribution of electrons in a molecule, leading to positive and negative ends on the molecule.

polarizer a filter that blocks light waves in all planes except one; a polarizer creates plane-polarized light.

polymer a very large molecule composed of repeating smaller units.

polymerization the process by which an organic compound reacts with itself to form a high-molecular-weight compound from repeating units of the original compound. Polymerization occurs by either cationic or free-radical mechanisms.

potential energy the energy a substance has due to its position or composition.

precursor the substance from which another compound is formed.

preparation a reaction in which a desired chemical is produced; for example, the dehydration of an alcohol is a preparation for an alkene.

primary carbocation a carbocation to which one alkyl group is bonded.

primary (1°) carbon a carbon atom that is attached to one other carbon atom.

principal quantum number (*n*) the first number in Schrödinger's electron wave equation, which tells the location of an orbital relative to the atom's nucleus.

priority rules the rules in the Cahn-Ingold-Prelog notational system that allow atoms or groups around a chiral carbon atom to be ranked by atomic weight.

product the substance that forms when reactants combine in a reaction.

projection a drawing of a molecule.

propagation step the step in a free-radical reaction in which both a product and energy are produced. The energy keeps the reaction going.

proton a positively charged particle in the nucleus of an atom.

protonation the addition of a proton (a hydrogen ion) to a molecule.

pure covalent bond a bond in which the shared electrons are equally available to both bonded atoms.

pyrolysis the application of high temperatures to a compound.

quantum mechanics the study of the mathematical formulas that describe the electronic structure of atoms.

quaternary carbon a carbon atom that is directly attached to four other carbon atoms.

racemate another name for "racemic mixture"; a 1:1 mixture of enantiomers.

racemic mixture a 1:1 mixture of enantiomers.

rate constant the proportionality constant of a reaction that reflects the concentration of reactants.

rate-determining step the step in a reaction's mechanism that requires the highest activation energy and is therefore the slowest.

rate equation the mathematical formula that relates the rate of a reaction to the concentration of the reactants.

rate of reaction the speed with which a reaction proceeds.

reactant a starting material.

reaction the process of converting reactants into products.

reaction energy the difference between the energy of the reactants and that of the products.

reagent the chemicals that ordinarily produce reaction products.

rearrangement reaction a reaction that causes the skeletal structure of the reactant to undergo change in converting to the product.

recrystallization a process based on solubility in which a substance is dissolved in a minimum amount of hot solvent, which is then cooled to allow new, purer crystals to form.

reduction the gaining of electrons by an atom or molecule. In organic compounds, a reduction is an increase in the number of hydrogen atoms in a molecule.

resolution to resolve; the process of separating enantiomers from a racemic mixture.

resonance the process by which a substituent either removes electrons from or gives electrons to a π bond in a molecule; a delocalization of electrical charge in a molecule.

resonance energy the difference in energy between the calculated energy content of a resonance structure and the actual energy content of the hybrid structure.

resonance hybrid the actual structure of a molecule that shows resonance. A resonance hybrid possesses the characteristics of all possible drawn structures (and consequently cannot be drawn). It is lower in energy than any structure that can be drawn for the molecule and thus more stable than any of them.

resonance structures various intermediate structures of one molecule that differ only in the location of the electrons.

R group see **alkyl group.**

ring-opening reaction a reaction that causes a cyclic structure to form a straight chain.

ring structure a molecule in which the end atoms have bonded, forming a ring rather than a straight chain.

rotation the ability of carbon atoms attached by single bonds to freely turn, which gives the molecule an infinite number of conformations.

R, S notational system see **Cahn-Ingold-Prelog notational system.**

saturated compound a compound containing all single bonds.

saturation the condition of a molecule containing the most atoms possible; a molecule made up of single bonds.

sawhorse projection a line drawing that is centered on two of the carbon atoms in a molecule and that shows, through perspective, the three-dimensional structure about them. Carbon atoms are represented by the intersection of bond lines. The arrangement resembles a carpenter's sawhorse.

secondary carbocation a carbocation to which two alkyl groups are bonded.

secondary (2°) carbon a carbon atom that is directly attached to two other carbon atoms.

separation technique a process by which products are isolated from each other and from impurities.

sequence rules the rules for establishing the order of atoms or groups in the Cahn-Ingold-Prelog notational system.

σ (sigma) antibonding molecular orbital a σ molecular orbital in which one or more of the electrons are less stable than when localized in the isolated atomic orbitals from which the molecular orbital was formed.

σ bond a bond formed by the linear combination of orbitals in such a way that the maximum electron density is along a line joining the two nuclei of the atoms.

σ (sigma) bonding molecular orbital a σ molecular orbital in which the electrons are more stable than when they are localized in the isolated atomic orbitals from which the molecular orbital was formed.

Simmons-Smith reaction the formation of a cyclopropane molecule via reaction of an alkene with iodomethylzinc iodide (ICH_2ZnI), the Simmons-Smith reagent.

skeletal structure the carbon backbone of a molecule.

sp hybrid orbital a molecular orbital created by the combination of wave functions of an s and a p orbital.

sp² hybrid orbital a molecular orbital created by the combination of wave functions of an s and two p orbitals.

sp³ hybrid orbital a molecular orbital created by the combination of wave functions of an s and three p orbitals.

spin quantum number (m_s) the fourth number in Schrödinger's electron wave equation, which tells the direction of spin on an electron.

staggered conformation one of the possible orientations of atoms or groups around two carbon atoms joined by a single bond. Atoms and groups bonded to the two carbons of a staggered conformer are positioned so there is maximum separation between them and therefore minimum interaction. (Compare with "eclipsed conformation.")

stereochemistry the study of the three-dimensional structure of molecules and how it affects their interactions.

stereogenic center a central atom that has four different atoms or groups bonded to it; also called **chiral center** or **asymmetric center.**

stereoisomers compounds with the same molecular formula but different arrangement of their atoms in three-dimensional space.

stereospecific the requirement of a specific stereochemical shape for a reaction to occur.

steric hindrance the blocking of one side of a molecule by a substituent so that any further bonding must occur on the opposite side. Steric hindrance causes the formation of *trans* stereoisomers.

straight-chain alkane a saturated hydrocarbon that has no carbon-containing side chains.

structural formula a chemical formula that shows not only the number and kind of atoms in a molecule but also their arrangement.

structural isomer also known as a *constitutional isomer,* structural isomers have the same molecular formula but different bonding arrangements among their atoms. For example, C_4H_{10} can be butane or 2-methylpropane, and C_4H_8 can be 1-butene or 2-butene.

subatomic particles a component of an atom; either a proton, neutron, or electron.

substituent group any atom or group that replaces a hydrogen atom on a hydrocarbon.

substitution the replacement of an atom or group bonded to a carbon atom with a second atom or group.

symmetry plane see **plane of symmetry.**

syn **addition** the addition of atoms to the same side of a molecule. (Compare with "*anti* addition.")

tautomers structural isomers that easily interconvert.

terminal alkyne an alkyne whose triple bond is located between the first and second carbon atoms of the chain.

terminal carbon the carbon atom on the end of a carbon chain.

termination step the step in a reaction mechanism that ends the reaction, often a reaction between two free radicals.

tertiary carbocation a carbocation to which three alkyl groups are bonded.

tertiary (3°) carbon a carbon atom that is directly attached to three other carbon atoms.

tetrahaloalkane an alkane that contains four halogen atoms on the carbon chain. The halogen atoms can be located on vicinal or nonvicinal carbon atoms.

tetrahedral angle an angle of 109°28", or approximately 110°. All of the bond angles in methane, CH_4, are tetrahedral angles.

thermodynamically controlled reaction a reaction in which conditions permit two or more products to form. The products are in an equilibrium condition, allowing the more stable product to predominate.

torsional strain strain caused by repulsion between groups in an eclipsed conformation.

***trans* isomer** a stereoisomer in which substituents are located on opposite sides of a double bond. (Compare with "*cis* isomer.")

transition state the point in a reaction at which the system has the most energy.

trigonal planar the shape of a molecule with an sp^2 hybrid orbital. In this arrangement, the σ bonds are located in a single plane separated by 60° angles.

triple bond a multiple bond composed of one σ bond and two π bonds. Rotation is not possible around a triple bond. Hydrocarbons that contain triple bonds are called alkynes.

unsaturated compound a compound that contains one or more multiple bonds; for example, alkenes and alkynes.

unsaturation refers to a molecule containing less than the maximum number of single bonds possible because of the presence of multiple bonds.

valence electrons the outermost electrons of an atom. The valence electrons of the carbon atom occupy the $2s$, $2p_x$, and $2p_y$ orbitals, for example.

valence shell the outermost electron orbit.

van der Waals forces the forces of attraction or repulsion between two unbonded groups or atoms due to unbalanced electrical forces.

van't Hoff rule predicts the maximum number of enantiomers that an optically active molecule can have; 2 raised to the nth power, where n equals the number of stereogenic centers.

vicinal a term that describes the location of two identical atoms or groups as being bonded to adjacent carbon atoms; a vicinal dihalide, for example. (Compare with "geminal.")

vinyl alcohol $CH_2 = CH - OH$

vinyl group the $CH_2 = CH -$ group.

wedge-and-dash projection a drawing of a molecule in which three types of lines are used to represent the three-dimensional structure of a molecule: 1) solid lines are bonds in the plane of the paper, 2) dashed lines are bonds extending away from the viewer, and 3) wedge-shaped lines are bonds oriented toward the viewer.

Wurtz reaction the coupling of two alkyl halide molecules to form an alkane.

X group "X" is often used as the abbreviation for a halogen substituent in the structural formula of an organic molecule.

Z **(zusammen)** the notation for the stereochemical arrangement in which the higher-ranked substituent groups are on the same side of the double bond. (See *E-Z* **notation.**)

Zaitsev rule states that the major product in the formation of alkenes by elimination reactions will be the more highly substituted alkene, or the alkene with more substituents on the carbon atoms of the double bond.

PERIODIC TABLE OF THE CHEMICAL ELEMENTS

Transition metals

IA									
1 **H** Hydrogen 1.01	IIA								
3 **Li** Lithium 6.94	4 **Be** Beryllium 9.01								
11 **Na** Sodium 22.99	12 **Mg** Magnesium 24.31								
19 **K** Potassium 39.10	20 **Ca** Calcium 40.08	21 **Sc** Scandium 44.96	22 **Ti** Titanium 47.90	23 **V** Vanadium 50.94	24 **Cr** Chromium 52.00	25 **Mn** Manganese 54.94	26 **Fe** Iron 55.85	27 **Co** Cobalt 58.93	
37 **Rb** Rubidium 85.47	38 **Sr** Strontium 87.62	39 **Y** Yttrium 88.91	40 **Zr** Zirconium 91.22	41 **Nb** Niobium 92.91	42 **Mo** Molybdenum 95.94	43 **Tc** Technetium (99)	44 **Ru** Ruthenium 101.07	45 **Rh** Rhodium 102.91	
55 **Cs** Cesium 132.91	56 **Ba** Barium 137.34	57 **La** Lanthanum 138.91	72 **Hf** Hafnium 178.49	73 **Ta** Tantalum 180.95	74 **W** Tungsten 183.85	75 **Re** Rhenium 186.21	76 **Os** Osmium 190.2	77 **Ir** Iridium 192.22	
87 **Fr** Francium (223)	88 **Ra** Radium (226)	89 **Ac** Actinium (227)	104 (261)	105 (262)	106 (263)	107 (262)	108 (265)	109 (266)	

Names not yet established for these elements

	58 **Ce** Cerium 140.12	59 **Pr** Praseodymium 140.91	60 **Nd** Neodymium 144.24	61 **Pm** Promethium (147)	62 **Sm** Samarium 150.35	63 **Eu** Europium 151.96
Lanthanides						
Actinides	90 **Th** Thorium (232)	91 **Pa** Protactinium (231)	92 **U** Uranium (238)	93 **Np** Neptunium (237)	94 **Pu** Plutonium (242)	95 **Am** Americium (243)

							VIIIA
							2 He Helium 4.00
		IIIA	IVA	VA	VIA	VIIA	
		5 B Boron 10.81	6 C Carbon 12.01	7 N Nitrogen 14.01	8 O Oxygen 16.00	9 F Fluorine 19.00	10 Ne Neon 20.18
		13 Al Aluminum 26.98	14 Si Silicon 28.09	15 P Phosphorus 30.97	16 S Sulfur 32.06	17 Cl Chlorine 35.45	18 Ar Argon 39.95

28 Ni Nickel 58.71	29 Cu Copper 63.55	30 Zn Zinc 65.38	31 Ga Gallium 69.72	32 Ge Germanium 72.59	33 As Arsenic 74.92	34 Se Selenium 78.96	35 Br Bromine 79.90	36 Kr Krypton 83.80
46 Pd Palladium 106.42	47 Ag Silver 107.87	48 Cd Cadmium 112.41	49 In Indium 114.82	50 Sn Tin 118.69	51 Sb Antimony 121.75	52 Te Tellurium 127.60	53 I Iodine 126.90	54 Xe Xenon 131.30
78 Pt Platinum 195.09	79 Au Gold 196.97	80 Hg Mercury 200.59	81 Tl Thallium 204.37	82 Pb Lead 207.19	83 Bi Bismuth 208.98	84 Po Polonium (210)	85 At Astatine (210)	86 Rn Radon (222)

64 Gd Gadolinium 157.25	65 Tb Terbium 158.93	66 Dy Dysprosium 162.50	67 Ho Holmium 164.93	68 Er Erbium 167.26	69 Tm Thulium 168.93	70 Yb Ytterbium 173.04	71 Lu Lutetium 174.97
96 Cm Curium (247)	97 Bk Berkelium (247)	98 Cf Californium (251)	99 Es Einsteinium (254)	100 Fm Fermium (257)	101 Md Mendelevium (258)	102 No Nobelium (259)	103 Lr Lawrencium (260)

ELECTRONEGATIVITY VALUES

H 2.1	

Li 1.0	**Be** 1.5		**B** 2.0	**C** 2.5	**N** 3.0	**O** 3.5	**F** 4.0
Na 0.9	**Mg** 1.2		**Al** 1.5	**Si** 1.8	**P** 2.1	**S** 2.5	**Cl** 3.0
K 0.8	**Ca** 1.0		**Ga** 1.6	**Ge** 1.8	**As** 2.0	**Se** 2.4	**Br** 2.8
Rb 0.8	**Sr** 1.0						**I** 2.4
Cs 0.7	**Ba** 0.9						

Many elements have been omitted to emphasize the basic pattern of electronegativity variation.